The Electrochemistry of Mars

Dust-Driven Processes, Biosignature Preservation, and In-Situ Resource Utilisation

Maria Perera-Burrell

Copyright

First edition
Published by: Maria Perera Burrell
Contact: maria@biooxynode.com

For my husband, Steven.

Author's Note

This book is not a review article and not a textbook. It is an argument. The distinction matters, because arguments require the reader to evaluate whether the reasoning holds, and I want to be clear from the start that this is what I am asking you to do.

The scientific framework developed here, that Mars has a spatially structured oxidation gradient driven by triboelectric atmospheric electrochemistry, and that this gradient is the primary unaccounted variable in both biosignature preservation and in-situ resource utilisation , is supported by published laboratory work, orbital observations, and flight data. It has not been tested by a direct field measurement on the Martian surface, because no instrument has yet been designed to make that measurement. The Mars Gradient Sampling System described in Chapter 32 is my attempt to specify what that instrument would look like. It is a concept, grounded in real hardware heritage and real physics, but it has not been built. I have tried to be precise throughout about the difference between what the evidence establishes, what the framework predicts, and what remains to be measured.

The engineering sections of this book, Parts Two and Three venture further from settled science than Part One does, and I want to acknowledge that directly. Technology readiness levels are presented as honest assessments, not promotional ones. Where I have assigned TRL 2 or TRL 3 to a system, I mean it: the physics is demonstrated, the engineering concept is coherent, but no integrated system exists and no Mars surface test has been run. The proposals in this book are directions for investment, not blueprints for immediate deployment.

Several of the ideas developed here emerged from conversations with colleagues whose names belong in any honest accounting of the book's origins. The connection between triboelectric discharge and perchlorate production as a spatially structured preservation variable was sharpened in discussion with researchers at the intersection of atmospheric chemistry and planetary geology whose published work I have cited throughout. The engineering architecture in Part Three draws on the published specifications and field test results of teams at Honeybee Robotics, JPL, the University of Antwerp, and the University of Lisbon, whose work I have tried to represent accurately and whose engineering judgements I have tried not to misrepresent. Any errors in my characterisation of their results are my own.

I began this project because I was frustrated by a gap in the literature, the gap between atmospheric physicists who study dust electrification and geochemists who study perchlorate distribution, between astrobiologists who assess ancient habitability and engineers who design resource extraction systems, between the scientists asking whether Mars was ever inhabited and the programme offices deciding where to send the next drill. That frustration is not a criticism of any individual. It is a structural observation about how planetary science organises itself, and this book is my attempt to build a bridge across that gap. Whether the bridge holds is for the reader, and ultimately for the data, to judge.

Maria Perera Burrell

Disclaimer

The scientific claims made in this book are the author's interpretations of published, peer-reviewed research and do not represent the official positions of any space agency, research institution, or funding body. Where the author's interpretive framework goes beyond what the cited literature directly establishes, this is indicated in the text by explicit qualification.

The engineering proposals described in Parts Two and Three including but not limited to the electrochemical ISRU architecture, the Mars Gradient Sampling System, and the phased demonstration ladder are conceptual designs grounded in publicly available hardware heritage data and published engineering literature. They are presented as research directions and design frameworks, not as validated systems, flight-qualified hardware, or proposals under active development by any organisation. Technology readiness level assessments are the author's own, based on the criteria defined by NASA's Technology Readiness Level scale and the evidence available at the time of writing.

Mass estimates, power budgets, and dimensional specifications cited in this book are preliminary and carry the uncertainties noted in the text. They are derived from publicly available sources including technical reports, conference proceedings, and peer-reviewed literature, and should not be relied upon for mission planning, procurement, or programme commitments without independent engineering verification.

The Mars Gradient Sampling System described in Chapter 32 is an original instrument concept proposed by the author. The individual component technologies referenced, including TRIDENT-heritage drilling systems, miniaturised field mills, and ExoMars-heritage Raman spectroscopy remain the intellectual property of their respective developers and are referenced for illustrative purposes only. No claim of ownership over those technologies is made or implied.

Nothing in this book constitutes engineering, scientific, or operational advice for any specific mission, programme, or system. The content is provided for informational and conceptual purposes only.

While every effort has been made to ensure accuracy, the author makes no representations or warranties, express or implied, regarding the completeness, accuracy, or applicability of the information presented. Scientific understanding evolves, and statements that are accurate at the time of writing may be superseded by subsequent measurement.

The author shall not be liable for any loss, damage, or consequence arising from the use or interpretation of the material contained in this book.

Preface

I had spent several years working at the intersection of atmospheric chemistry and planetary science, and I kept encountering the same assumption in the literature: that the surface of Mars is essentially static. That the chemistry we measure in Martian rocks today reflects conditions that prevailed billions of years ago, preserved more or less intact until a rover drills into them. That the job of Mars exploration is archaeological ,find the right ancient environment, read the record it contains, determine whether life was ever part of it.

The frustration was not that this view was wrong in its broad strokes. Mars did lose its oceans and most of its atmosphere. Its ancient mineralogy does record past conditions. But the assumption of stasis papered over something that the data kept insisting on: the surface of Mars is being chemically processed right now, by mechanisms that are well understood, at rates that matter for what any mission will actually find when it drills. The atmosphere is manufacturing oxidants through triboelectric discharge in dust devils, while radiation-driven processes simultaneously modify and degrade those products. Those oxidants are diffusing into exposed rock. The ancient record, in other words, is not just old ,it is being edited by an ongoing process that varies in intensity across the planet in ways that no current mission has been designed to map.

Once I saw that, I could not unsee it. The implications run from basic science ,which sites preserve biosignatures and which do not ,through mission design ,how deep do you need to drill, and where ,to the engineering of a human settlement that must be built on soil running at perchlorate concentrations ninety times the residential safety threshold. These are not separate questions. They are expressions of the same underlying system, and understanding that system changes how you approach all of them.

This book is an attempt to assemble that argument coherently, to be honest about where the evidence is strong and where it is not, and to follow the implications wherever they lead ,including into engineering territory that is sometimes left out of planetary science writing. It is written for scientists, mission planners, and engineers who are thinking seriously about Mars, and who want a framework that accounts for the planet as it actually is rather than as the conventional narrative assumes it to be.

The science in these pages is grounded in published, peer-reviewed literature. Where claims are speculative or predictions untested, they are flagged as such. The engineering proposals are assessed at their actual technology readiness levels, not their aspirational ones. Where I have been wrong, I expect the data to show it ,which is, in the end, the only honest position any framework can occupy.

Introduction: Why This Argument, Why Now

Mars exploration has entered an unusual moment. The scientific case for ancient habitability is stronger than it has ever been. The Curiosity rover confirmed organic molecules in ancient mudstones. The Perseverance rover is caching samples from a river delta that deposited sediment into a lake roughly three and a half billion years ago. Orbital spectroscopy has mapped clay minerals, carbonates, and sulfates across ancient terrains ,the mineralogical fingerprints of prolonged liquid water. We are not speculating about whether Mars once had conditions compatible with life. We have the geological evidence in front of us.

And yet the question of whether Mars ever hosted life remains unanswered, and the missions designed to answer it have returned ambiguous or negative biosignature results from sites that every prior criterion said should be excellent targets. This is not necessarily because Mars was never inhabited. It may partly be because the framework guiding exploration is missing a variable ,one that, once included, changes not just where we should look but how deeply, and what we should expect to find when we get there.

That variable is the modern chemical activity of the Martian surface.

In 2008, the Phoenix lander detected perchlorate in polar soil at concentrations around half a percent by weight. Perchlorate is a powerful oxidizer that, at those concentrations, is capable of destroying organic molecules. When the Curiosity rover measured comparable perchlorate concentrations thousands of kilometres away in equatorial Gale Crater, the global distribution became undeniable.Something is producing perchlorate everywhere on Mars, continuously and efficiently enough to maintain that concentration in the presence of competing destruction processes. Laboratory work conducted through the 2010s identified the mechanism: triboelectric discharge in dust devils converts chloride minerals to perchlorate through plasma chemistry driven by the electrical breakdown of the thin Martian atmosphere. In October 2024, the SuperCam microphone aboard Perseverance confirmed the mechanism is not laboratory inference ,it recorded fifty-five discrete electrical discharge events inside Martian dust devils and storm fronts, establishing beyond reasonable doubt that this chemistry is happening now, on the surface of Mars, at scales observable by a rover instrument.

The consequence of this for the ancient record is specific and consequential. Every rock exposed at the Martian surface has been in contact with a chemically active atmosphere for as long as it has been exposed. The perchlorate-producing, oxidant-generating atmospheric chemistry that dust devils drive has been working on whatever sits at the surface since erosion first uncovered it. An ancient lake sediment exposed for five hundred million years in a high-activity dust devil zone has received a very different cumulative oxidative dose than the same sediment exposed in a quiet region for fifty million years. The organic molecules, isotopic signatures, and mineral assemblages that might record past life ,the biosignatures exploration is looking for ,have been subject to that ongoing processing in ways that vary across the planet and with depth below the surface.

This book develops that argument and follows it through. Part One establishes the mechanism, the observational evidence for its spatial structure across Mars, and the implications for biosignature preservation and site selection. The core claim of Part One is that Mars has a preservation gradient ,a spatially structured pattern in how much modern atmospheric chemistry has modified the ancient record at any given location ,and that this gradient should be a primary variable in mission planning rather than an afterthought. This is presented as a well-motivated hypothesis grounded in established physics and chemistry, not as settled fact. The specific tests that would confirm or refute it are identified explicitly.

Part Two extends the framework into resource utilization and settlement planning. The same atmospheric processes that complicate biosignature preservation also produce materials with genuine engineering value ,perchlorate as an oxygen feedstock, hygroscopic salts as water-concentrating agents, electrostatically charged dust as a sortable mineral stream. Understanding Mars as an active chemical system that has been processing materials for billions of years is the foundation for designing resource extraction that works with what the planet provides rather than importing complete industrial solutions from Earth. Part Two is realistic about what is established, what is demonstrated only at laboratory scale, and what remains conceptual.

Part Three is the engineering section. It begins with what flight hardware has actually confirmed ,MOXIE's TRL-9 demonstration of solid oxide electrolysis on Mars, the SuperCam electrical discharge recordings, the RedWater subsurface ice drilling demonstrations on Earth ,and builds toward an integrated architecture for early human surface operations. It addresses failure modes explicitly, proposes a phased demonstration ladder from orbital characterization through robotic precursors to crewed missions, and argues that site selection for human missions is an engineering decision that the dust activity framework of Part One directly informs. It also addresses gaps that engineering discussions of Mars habitability typically avoid: the perchlorate toxicology problem for crew health, the electrostatic discharge hazard to sensitive hardware, and the implications of oxidation penetration depth for sample return curation.

Different readers will find their entry point in different sections. Scientists and astrobiologists will find the core argument in Chapters 1 through 13. Mission planners will find the operational implications in Chapters 8, 13, and 28 through 30. Engineers designing early human Mars systems will find the resource architecture in Chapters 14 through 27. All three groups will find the framework incomplete in specific ways that are named rather than concealed, because the most useful contribution this book can make is not to provide final answers but to identify the right questions and specify what measurements would resolve them.

One word about scope. This book does not address Mars terraforming, interplanetary economics, or the long-term political future of Mars settlement. It is not a general Mars primer. It assumes the reader has basic familiarity with Mars exploration history and planetary science concepts, though technical terms are explained when they first appear. The argument is focused: Mars is an active chemical system, that activity has structured the ancient record in ways that exploration has not yet fully accounted for, and getting that accounting right matters both for the science and for the engineering of eventual human presence. Everything here is in service of making that case as clearly and honestly as the available evidence allows.

Mars is still there. The questions are still open. This book is an attempt to sharpen the tools we use to approach them.

Part One: The Active System
Electrochemistry, Preservation, and the Science of a Dynamic Mars

Chapter 1: A Planet Misunderstood

The standard narrative about Mars runs something like this: the planet wound down billions of years ago, losing its magnetic field, its atmosphere, and eventually its liquid water, leaving behind a cold desert preserved almost exactly as it was during the Hesperian period. What the rovers photograph today, iron-stained plains, eroded craters, dust, reflects conditions that stabilized when the last surface water evaporated or froze. The chemistry locked into Martian rocks is ancient chemistry. The task of exploration, under this view, is essentially archaeological: find the right sediments, read the record they contain, and determine whether life was ever part of that record.

That narrative carries real scientific weight. Mars did lose most of its atmosphere. It did transition from a wetter, warmer state to the frozen desert we observe now. The ancient geological record is genuinely there, encoded in clay minerals, sulfate deposits, and carbonate formations that formed under conditions very different from the present. None of this is in dispute.

What the narrative misses is what has been happening to that record ever since it formed.

In 2008, NASA's Phoenix lander set down near the Martian north pole and sampled the soil with an instrument called the Thermal and Evolved Gas Analyzer. Among its findings was something unexpected: perchlorate, a chlorine-oxygen compound, present at concentrations around half a percent by weight. Perchlorate is a powerful oxidizer. It also dramatically lowers the freezing point of water, enabling liquid brines at temperatures far below what pure water could tolerate. The discovery drew immediate attention as a contamination hazard for future human missions, since perchlorate interferes with thyroid function in mammals, but its scientific significance ran deeper. The question was where it came from and what it meant that it was there at all (Hecht et al., 2009).

If perchlorate had formed through ancient ultraviolet photochemistry and simply accumulated over geological time, its distribution might be expected to reflect that ancient history, patchy, tied to specific geological contexts, perhaps concentrated in certain sedimentary layers. Instead, when the Curiosity rover began sampling in Gale Crater years later, more than four thousand kilometres from the Phoenix landing site and in a geologically distinct equatorial setting, it found perchlorate at comparable concentrations. The two sites shared almost nothing in terms of ancient geology. They shared perchlorate nonetheless. The implication of a globally distributed oxidizer, present at roughly similar concentrations across vastly different terrains, was that something was producing it continuously and distributing it widely, not a relic of the ancient past but an ongoing output of a process still operating (Glavin et al., 2013).

Laboratory work conducted through the 2010s pointed toward a mechanism. When Martian soil simulants containing chloride minerals are agitated inside low-pressure chambers designed to approximate Mars atmospheric conditions, triboelectric charging occurs as particles collide and exchange electrons. Because the Martian atmosphere is so thin, surface pressure averages around six to seven millibars, less than one percent of Earth's sea-level pressure, there are far fewer ions available to dissipate the charge that builds up. The result is that particles in a Martian dust devil accumulate electrical charge far more efficiently than particles in a terrestrial dust storm. When those fields grow strong enough, electrical breakdown occurs: brief plasma discharges form, and within those discharges, chloride minerals can be oxidized stepwise through chlorate to perchlorate. Laboratory measurements found this triboelectric pathway produces perchlorate at rates orders of magnitude higher than ultraviolet photochemistry under the same Mars-relevant conditions (Atreya et al., 2011; Jackson et al., 2016). The mechanism is consistent with what we observe, and it is not ancient. It operates in every dust devil crossing the Martian surface today.

This is not a minor refinement to our picture of Mars. It changes the interpretive framework in a specific and consequential way.

Consider what it means for a rock. Suppose a sedimentary formation deposited in a Martian lake basin roughly three and a half billion years ago was buried, lithified, and later exhumed by erosion. From the moment it became exposed at the surface, that rock has been in contact with a chemically active atmosphere generating oxidizing species. Over hundreds of millions of years of surface exposure, oxidants produced by dust activity can diffuse into the pore spaces of the rock, react with any organic molecules present, alter mineral assemblages toward more oxidized phases, and overprint isotopic signatures that might otherwise record ancient biological or abiotic processes. The rock is not sitting inert. It is being slowly processed by a system that has been running continuously throughout its exposure history.

The depth to which this processing penetrates depends on local conditions, the permeability of the rock, the concentration gradient of oxidizing species at the surface, and the cumulative duration of exposure. Laboratory studies suggest that oxidants produced by electrical discharge can work their way millimetres to centimetres into rock samples over experimental timescales, and scaling those rates to geological exposure times produces estimated alteration depths ranging from centimetres to perhaps a meter or more, depending on conditions. Below that altered zone, ancient chemistry is likely to be better preserved. Above it, the record has been modified to a degree that varies with how long the material has been exposed and how chemically active its particular location has been.

This depth-dependent preservation is not unusual in geology. On Earth, exposed rock faces show weathering profiles where fresh unaltered material at depth gives way to progressively more altered rock approaching the surface, products of interaction with atmospheric oxygen, liquid water, and biological activity. The relevant chemistries on Mars differ, but the underlying principle is the same. Exposure at a planetary surface means exposure to the atmosphere acting on that surface, and on Mars that atmosphere is electrochemically active. Ancient materials record ancient conditions only insofar as subsequent atmospheric interaction has not overprinted them.

What makes this particularly significant for Mars exploration is that the rate of atmospheric processing is not uniform across the planet. Dust devil activity, the primary driver of the electrochemical pathway described above, varies substantially by region, season, and local topography. Surveys of Martian dust devil tracks and active vortices using images from the Mars Reconnaissance Orbiter's Context Camera have documented that some regions are highly active while others see far less convective dust lifting. The differences are not random but reflect underlying controls: surface thermal properties, topography, latitude, and seasonal heating patterns. Where dust activity is high, atmospheric processing should, over time, deliver a greater cumulative oxidative dose to exposed surface materials. Where activity is low, that cumulative dose will be smaller, and materials at or near the surface should retain more of their original chemistry.

This suggests the possibility of a preservation gradient across Mars, a spatial structure in how much modern atmospheric chemistry has modified the ancient record at any given location. It is important to be precise about the status of this idea. The individual components are well-supported: the triboelectric mechanism in the laboratory, the spatial structure of dust activity from orbital surveys, the demonstrated oxidative effect of perchlorate on organics under Mars-relevant conditions. What has not yet been measured directly is whether sites with contrasting dust activity histories show correspondingly different degrees of organic preservation or surface oxidation. That is the specific test the framework predicts, and it has not yet been carried out in a targeted way. The preservation gradient is a well-motivated hypothesis grounded in established physical and chemical processes, not an established fact. That distinction matters and will be returned to throughout this book.

The practical significance of the hypothesis is this: choosing where to look for ancient biosignatures on Mars is not only a question of where ancient habitability was greatest. It is also a question of where ancient evidence has been least disturbed since. A site with excellent ancient geology, a well-preserved lake delta with clay-rich sediments indicating prolonged aqueous chemistry, but a long history of intense dust activity may have had its organic record substantially reduced by cumulative oxidative processing. A site with more modest ancient features but a quieter atmospheric history may have retained more of whatever chemical record it once held. Mission planning that accounts only for ancient habitability while treating modern preservation conditions as uniform is working with an incomplete picture of what will actually be found.

The evidence that supports this view and the gaps that remain in it are what the subsequent chapters work through. The mechanism of dust-driven electrochemistry is examined in detail first, because it is the physical foundation on which the rest of the argument rests. Then the observational evidence for spatial structure in dust activity, the question of oxidation penetration depths, the implications for biosignature science, and the practical consequences for exploration planning are developed in turn. Later chapters extend the framework into resource extraction and settlement planning, where the same atmospheric processes that complicate preservation also produce materials with potential engineering value. Throughout, the aim is to be clear about what the data show, what the models predict, and where the two have not yet been compared in ways that would allow the predictions to be confirmed or revised.

Chapter 2: The Mechanics of Dust-Driven Chemistry

Understanding how a dust devil produces perchlorate requires following a chain of physical processes from atmospheric convection down to chemical reaction. Each link in that chain has been studied, in the laboratory, in terrestrial field measurements, or through direct observation on Mars. The full pathway is not characterized in every detail, and meaningful questions remain about how efficiently it operates at planetary scale. But the core mechanism is well-established, and working through it carefully is necessary because it forms the physical foundation for much of what follows.

Triboelectric charging is a well-understood phenomenon: when two materials make contact and then separate, electrons transfer between their surfaces based on differences in electron affinity. One material retains excess electrons and becomes negatively charged; the other loses electrons and becomes positively charged. At the scale of individual dust grains in a turbulent vortex, the same physics operates with each collision, transferring small quantities of charge that add up across millions of particle interactions over the lifetime of the vortex.

The efficiency of this charging depends strongly on atmospheric conditions, and here Mars differs from Earth in a directly relevant way. Earth's atmosphere, dense enough to carry abundant ions and water vapor, allows charge to dissipate from particle surfaces relatively quickly, the atmosphere itself conducts charge away before it can accumulate to high levels. The Martian atmosphere provides no equivalent dissipation. Surface pressure averages around six to seven millibars, less than one percent of Earth's sea-level value, and atmospheric water vapor is trace-level. Charge carriers are sparse, atmospheric electrical conductivity is low, and particles hold their charge far longer after each collision. Laboratory experiments using

Martian soil simulants in low-pressure chambers have measured charge accumulation under Mars-relevant conditions at levels roughly ten times or more above what the same experiments produced at Earth atmospheric pressure (Krauss et al., 2003; Kok and Renno, 2009). Inside a Martian dust devil, charge separation proceeds efficiently.

Inside the vortex, this charging is also spatially sorted. Smaller, finer particles preferentially acquire negative charge in contacts with larger particles, while coarser material tends to acquire positive charge, a pattern whose specifics depend on mineral composition and surface conditions but which is broadly consistent across experimental work. Since dust devils naturally segregate particles by size, fine dust drawn up into the central rising column, coarser material lofted less efficiently and concentrated lower, the compositional sorting and the charge sorting reinforce each other. Fine, negatively charged particles move upward; coarser, positively charged material remains at lower levels. The result is a vertically oriented charge separation that generates electric fields directed roughly along the axis of the vortex. Direct measurements of electric fields inside terrestrial dust devils have recorded values of several thousand volts per meter, with a geometry consistent with this separation pattern (Renno et al., 2004; Farrell et al., 2004).

When field strength exceeds the Paschen breakdown threshold for the ambient gas, the gas can no longer sustain it. A conducting channel forms momentarily through partial ionization, and the field collapses. The Paschen threshold depends on gas pressure and composition: in a low-pressure environment like Mars, breakdown occurs at lower absolute field strengths than at Earth's surface. Once breakdown happens, the discharge channel briefly becomes a partial plasma, a region of energetic electrons, ions, excited molecular species, and reactive radicals including oxidized forms of both oxygen and chlorine derived from the gas phase and from mineral surfaces in contact with the discharge region.

It is within these brief plasma events that the perchlorate chemistry becomes possible. Chloride minerals, primarily sodium, magnesium, and calcium chlorides, are widely distributed in Martian regolith. When chloride-bearing particles are exposed to plasma conditions, chlorine can be liberated and enter a sequence of oxidation reactions driven by reactive oxygen species. The pathway proceeds from chloride through hypochlorite and chlorite intermediates, then to chlorate, and finally to perchlorate as chlorine reaches its maximum oxidation state. The reaction sequence involves numerous short-lived intermediates and is not simple, but the net transformation, chloride to perchlorate through successive oxidation steps energized by electrical discharge, has been reproduced in laboratory experiments conducted under Mars-relevant gas composition and pressure (Carrier and Kounaves, 2015).

The efficiency comparison with alternative production pathways is important for understanding why this mechanism matters. Ultraviolet photochemistry, previously the favoured explanation for Martian perchlorate, can drive oxidation reactions in the atmosphere but operates slowly. When Atreya and colleagues (2011) compared production rates from the triboelectric pathway against UV photochemistry under equivalent Mars conditions, the discharge mechanism produced perchlorate at rates several orders of magnitude higher. UV photochemistry may contribute at some level, and the relative contribution of each pathway at planetary scale is still an active area of research, the dominance of triboelectric discharge across all production routes is supported by laboratory evidence but has not been confirmed through field measurements that quantify actual fluxes. What the laboratory data establish firmly is that triboelectric discharge is a far more efficient known mechanism under Mars conditions, and that its production rates are consistent with the surface concentrations Phoenix and Curiosity measured, given plausible assumptions about dust activity integrated over geological time.

Once produced, these chemical products do not stay where they are made. Fine particles processed in the plasma region of a dust devil are carried aloft by the vortex's updraft and eventually settle out downwind, potentially hundreds of kilometres from their origin. During planet-wide dust storms, which Mars experiences regularly and which can keep dust suspended globally for weeks, transport distances can extend to the full planetary scale. This redistribution contributes to the global uniformity in perchlorate concentrations observed across the Phoenix and Curiosity sites. It also means that surface chemistry at any given location reflects a mixture of local production and long-range deposition, and disentangling these contributions requires systematic spatial datasets that are still being developed.

One aspect of dust devil dynamics worth examining more closely is the scale of influence each vortex exerts on its surroundings. The visible dust column represents only the loaded core of the system. Field campaigns deploying dense networks of weather stations around terrestrial dust devils have shown that the dynamic influence extends substantially further. In documented cases, a dust devil with a visible diameter of roughly nineteen meters produced measurable pressure perturbations at stations nearly two hundred meters away, with the surface inflow pattern, air being drawn horizontally toward the vortex base, extending over an area roughly ten times the visible radius in each direction (Kurgansky et al., 2016). The dynamically active footprint, where surface material is being entrained and the atmosphere is being processed, is therefore on the order of one hundred times larger than the visible cross-section suggests. Applied to Mars, where vortices frequently reach diameters of hundreds of meters and occasionally exceed a kilometre, even moderate detection densities from orbital surveys could translate to a large fraction of the surface in active regions being periodically swept by the influence of nearby vortices.

The spatial distribution of dust devil activity across Mars has been characterized through systematic analysis of Context Camera imagery over multiple Martian years. Dust devil tracks, dark streaks where a vortex has cleared bright surface dust and exposed darker material beneath, accumulate as a record of where activity has been concentrated. Active vortices directly imaged in Context Camera frames add size and temporal information. Together, these datasets show clear geographic structure: regions with low thermal inertia surfaces that heat and cool rapidly during the Martian day produce more convective instability and more dust devil activity. Certain topographic configurations Favor vortex initiation. Lower latitudes receive more solar energy per unit area and generate stronger surface heating. The result is a planet where some regions are substantially more active than others, and where that pattern repeats each Martian year driven by the same underlying physical factors.

The connection from this activity structure to surface chemistry follows from the mechanism described above. If perchlorate production scales with vortex frequency and intensity, as laboratory work suggests it should, then regions with higher dust devil densities should accumulate perchlorate at higher rates and deliver a greater cumulative oxidative dose to exposed surface materials over time. The magnitude of resulting compositional differences between high-activity and low-activity regions is the critical open question. If the differences are large, they have direct implications for where biosignature material is most likely to have survived in accessible form. If they are small, the effect on exploration strategy is correspondingly limited. This is precisely the kind of question that targeted multi-site measurements, comparing chemistry at locations specifically chosen to span the range of documented activity levels, could answer. Current mission planning has not yet incorporated this as a deliberate objective.

What the existing body of evidence does establish is the physical coherence of the mechanism and the reality of its primary input variable. The triboelectric pathway from chloride to perchlorate works under Mars conditions in the laboratory. The atmosphere provides the energy and the dust provides the feedstock whenever convection is active. Dust activity is spatially structured in ways that orbital data now document reasonably well. The framework connecting these established components to preservation gradients in surface chemistry is logically sound and consistent with everything known about the relevant physical and chemical processes. What it awaits is direct field confirmation, measurements designed to test its predictions rather than measurements designed for other purposes that happen to bear on it incidentally. How to design those measurements, and what they would tell us, is the subject of later chapters.

Chapter 3: Mapping Active Chemistry

Arguing that Mars is a chemically active world is one thing. Showing where that activity is concentrated, how intense it is, and whether it leaves measurable traces in surface composition is another. The argument depends on data, and the data come from several independent sources that do not all point at the same things. Orbital imaging covers the whole planet but measures dust movement, not chemistry. Surface missions measure chemistry in detail but at single points. Spectroscopy from orbit maps mineral distributions but with a targeting strategy that introduces its own biases. Climate models predict where dust activity should be favoured but need ground truth to validate. None of these datasets alone answers the question of how spatially structured chemical processing is across Mars. Together, and with appropriate attention to what each can and cannot show, they begin to build a coherent picture.

The most direct observational record of dust devil activity comes from the Context Camera aboard the Mars Reconnaissance Orbiter, which has been imaging Mars continuously since 2006 at roughly six-meter resolution. The camera was designed primarily to provide geographic context for the orbiter's higher-resolution instruments, but its combination of broad aerial coverage and sufficient resolution to detect meter-scale features makes it well-suited for systematic vortex surveys. A dust devil in a Context Camera frame appears as a bright or dark column, sometimes with a visible shadow that confirms its three-dimensional structure, often with a dark track behind it where the vortex has swept surface dust away and exposed the darker substrate beneath. These tracks accumulate over time as a spatial record of where vortex activity has been concentrated. Active vortices imaged in real time add information about size and frequency.

Recent analysis applied machine learning classification to more than one hundred thirty thousand Context Camera images spanning multiple Martian years, identifying thousands of active dust devils across the dataset. The geographic distribution that emerges from this survey is not uniform. Activity clusters at certain latitudes and longitudes, peaks during specific seasonal windows corresponding to maximum solar heating and the deepest convective boundary layers, and largely avoids other regions and seasons. The pattern is coherent and physically interpretable: high-activity zones correspond to surfaces with low thermal inertia, fine dust and sand rather than rock, that heat rapidly during the Martian day and build steep temperature gradients in the lower atmosphere. Topographic configurations that enhance local wind shear or concentrate surface heating appear in the data as elevated-activity corridors. This is what the thermodynamics of convective vortex formation predicts, and the data bear it out.

Surface measurements from the Curiosity rover add an important layer of validation. Curiosity's meteorological suite records pressure, wind, temperature, and humidity continuously, and passing vortices register clearly as sharp pressure drops lasting seconds to minutes. Over years of operation in Gale Crater, the rover has logged thousands of vortex encounters. When encounter rates are compared against a dust devil activity index calculated from measured atmospheric variables, a thermodynamic metric that combines sensible heat flux and convective efficiency, the correlation is strong. Active periods in the index predict elevated encounter rates; quiet periods predict suppression. Individual encounters remain stochastic, but the aggregate follows the model prediction closely enough to validate the thermodynamic framework as a reliable predictor of activity intensity. That validation matters because it means orbital thermal and meteorological data can be used to forecast dust devil activity at locations where no surface mission has operated, extending the activity characterization beyond the handful of places where rovers have actually been.

What these datasets establish with reasonable confidence is that dust devil activity on Mars is spatially structured and that the structure is physically explained. What they do not yet provide is the next link in the chain: a direct comparison between dust activity levels and surface chemical composition at matched locations. The perchlorate measurements that exist, from Phoenix at high northern latitudes and from Curiosity in equatorial Gale Crater, both come from sites that orbital surveys characterize as regions of moderate dust devil activity. Neither site represents the high-activity extreme, where the framework predicts the greatest cumulative oxidative processing, nor the low-activity extreme, where the best preservation of ancient chemistry would be expected. The dataset as it currently stands cannot test the core prediction because it does not span the range of conditions the prediction involves. This is an honest and important limitation. The global distribution of perchlorate, consistent concentrations across two geologically unrelated sites thousands of kilometres apart, supports a continuous global production mechanism, which is consistent with dust-driven electrochemistry. It does not confirm that production rate scales spatially with dust activity, because that specific comparison has not been made.

The Compact Reconnaissance Imaging Spectrometer for Mars, also aboard the Mars Reconnaissance Orbiter, offers a different view of surface chemistry through visible and near-infrared spectroscopy. It identifies mineral phases by their characteristic reflectance signatures, mapping distributions of phyllosilicates, sulphates, carbonates, iron oxides, and other mineral classes across the planet's surface. If sustained oxidative processing from dust-driven chemistry alters surface mineralogy over geological time, driving mineral assemblages toward more oxidized phases, destroying hydrated minerals in exposed layers, or depositing oxidized dust, one might expect spectroscopic differences between regions with long histories of intense dust activity versus regions that have been relatively quiet. Whether that signal is detectable depends on how large the mineralogical change actually is, which is not yet known.

Testing this using existing spectroscopic data runs into a structural problem. The Compact Reconnaissance Imaging Spectrometer's observation archive is not a uniform survey of the Martian surface. It is a targeted dataset, with high-resolution observations concentrated at locations identified as geologically interesting from lower-resolution context mapping or from knowledge of ancient aqueous activity. Ancient lake beds, hydrothermal systems, impact crater floors, and valley networks receive disproportionately dense coverage. Typical plains terrain and regions with no compelling ancient geological interest receive far less. The practical consequence is that the archive overrepresents precisely the kinds of terrains that Mars science has prioritized on other grounds, terrains that may or may not correspond to the regions of highest or lowest dust activity. Attempting a spatial correlation between spectroscopic mineral maps and dust activity maps from Context Camera data using this archive requires accounting for that selection bias carefully, or the result will reflect the targeting choices of the spectroscopy team as much as any real physical relationship in the data.

This kind of cross-dataset spatial correlation, systematically comparing where dust activity is high with what the mineralogy looks like at those locations, and doing the same for low-activity regions, has not yet been performed comprehensively. The dust activity mapping from Context Camera analysis has only recently reached sufficient spatial coverage and temporal baseline to support it. The analysis is technically tractable: it would require co-registering the geographic coordinates of spectroscopic observations with dust devil detection densities from the Context Camera survey and computing statistical measures of whether mineralogical variation correlates with activity variation across locations where both datasets have coverage. The results would be informative in either direction. A positive correlation between high dust activity and more oxidized surface mineralogy would be consistent with the processing framework developed here.

No correlation, or a correlation in the wrong direction, would suggest that either the effect is too small to be detected in existing mineralogical maps, or that ancient geological variation dominates modern atmospheric processing in terms of observable mineral signatures, or that the mechanism operates differently than the framework predicts. All of these outcomes would be scientifically useful.

It is worth being explicit about the logical structure of the hypothesis and what would be required to confirm or refute it. The preservation gradient idea rests on several components that each need to hold. Dust devil activity must be spatially structured rather than uniform, the Context Camera surveys confirm this. The chemical products of that activity, particularly perchlorate and other oxidizing species, must affect organic preservation and mineral stability in ways relevant to biosignature science; laboratory work supports this at the sample level. The timescales of processing must be long enough for cumulative effects to be detectable, given that surface exposures on Mars can span hundreds of millions to billions of years and the mechanism operates continuously, this condition seems likely to be met. And the differences in processing rate between high-activity and low-activity regions must be large enough to create compositional gradients that matter for exploration decisions, this is the component that field measurements have not yet tested.

If all four components hold at the levels the framework assumes, then the surface of Mars carries a chemical record that superimposes ancient geological history and modern atmospheric processing in a way that varies spatially. Disentangling those two signals at any given site requires knowing something about both the ancient depositional environment and the subsequent exposure and processing history. Depth profiling, sampling at multiple depths to compare altered surface material with less-altered subsurface material, provides one tool for separating these contributions. Comparing multiple sites with similar ancient geology but different modern activity histories provide another. Neither approach has been applied systematically, because it requires mission designs that treat modern atmospheric chemistry as a first-order variable in site selection alongside ancient habitability indicators. Current practice treats modern chemistry as a hazard to be accounted for rather than a variable to be characterized and used to guide where and how to sample.

The practical direction this points toward is not complicated in concept, though it is demanding in terms of mission coordination. A campaign targeting sites specifically selected to span the full documented range of dust activity, from the most active regions identified in Context Camera surveys to the quietest accessible terrains, with similar enough ancient geology to control for that variable, would generate the comparison dataset needed to test whether compositional differences scale with activity as predicted. Sample return from multiple such sites, or even in-situ measurements with sufficient analytical capability to detect organic content and oxidation state profiles with depth, would be informative. The science case for including activity characterization as a site selection criterion in future landed missions is straightforward once the framework is laid out, but it requires making that criterion explicit in mission requirements rather than leaving it as an afterthought.

The datasets available today support the first step of the argument: dust activity is real, spatially structured, physically explained, and persistent across Martian years. The second step, that this activity leaves measurable chemical signatures that vary in the same spatial pattern, is the prediction the framework makes and the question that future coordinated measurements can answer. Getting that answer is not a marginal scientific question. It goes directly to whether the sites Mars exploration has been prioritizing are the ones most likely to yield the biosignature evidence the missions are designed to find.

Chapter 4: The Chemistry of Ancient Life

Searching for biosignatures on Mars is not simply a matter of pointing instruments at ancient rocks and reading off the result. The challenge has two distinct components, and conflating them leads to unrealistic expectations about what any given mission can demonstrate. The first component is analytical: Mars missions operate under mass, power, and bandwidth constraints that force genuine compromises in the instruments carried. The second component is interpretive: even with perfect instruments, biosignatures from billion-year-old rocks are ambiguous by nature, because the chemical and morphological features that life produces on Earth can also be generated, at least in part, by purely geological processes. Understanding both components is necessary before considering how modern oxidative chemistry complicates the picture further.

On Earth, a researcher studying candidate biosignatures in ancient Australian or South African charts has access to essentially unlimited analytical power. Samples can be cut, polished, stained, and examined by electron microscopy. Isotope ratios can be measured at micrometer spatial scales by ion microprobe. Organic compounds can be extracted and subjected to gas chromatography, mass spectrometry, and compound-specific isotope analysis. If one technique is ambiguous, another can be applied. Time and funding permitting, the same sample can be examined by a dozen different methods until the evidence either converges on a credible biological interpretation or fails to support one. The process is iterative, and negative results shape the approach as much as positive ones.

Mars exploration cannot work this way, at least not yet. Every kilogram of scientific payload on a landed mission is purchased at enormous cost in propellant, structural mass, and engineering complexity. Instruments designed for Mars must survive months of interplanetary transit, entry heating, landing shock, and then years of operation in a dusty, cold, radiation-exposed environment while receiving commands with round-trip light-time delays of up to forty-four minutes. These constraints are not engineering failures; they reflect the genuine physics of sending sophisticated equipment across interplanetary distances. But they mean that the instruments reaching Mars represent chosen priorities, not comprehensive capability.

The Curiosity rover's Sample Analysis at Mars suite pyrolyzes rock and soil samples and analyzes the released gases by gas chromatography and mass spectrometry. It has detected organic molecules in multiple sample types, including thiophenic, aromatic, and aliphatic compounds in ancient mudstone at Gale Crater, a significant finding, as these materials have survived in detectable quantities despite the hostile surface environment. What Sample Analysis at Mars cannot do is determine the three-dimensional molecular structure of complex organics, distinguish between all possible structural isomers, or measure spatial distributions within a rock at fine scales. The Perseverance rover's SHERLOC instrument uses deep-ultraviolet Raman and fluorescence spectroscopy to map organic compounds and minerals at submillimetre scales, providing spatial context that Curiosity's pyrolysis approach lacks. Together these instruments represent genuine advances. They also represent the gap between in-situ planetary science and what a well-equipped terrestrial laboratory can do with the same sample.

This matters because identifying a biosignature with confidence requires not just detecting the compound or feature of interest but demonstrating that biological explanation is more parsimonious than any available abiotic explanation. That demonstration depends on the weight and independence of multiple lines of evidence, and the instrumental limitations of Mars missions constrain how many independent lines can be measured at the same location.

The categories of potential biosignatures, molecular, morphological, isotopic, and mineralogical, each carry their own strengths and limitations as evidence of past life. Morphological biosignatures, such as preserved cell walls or stromatolitic layering, are visually compelling but notoriously vulnerable to misinterpretation. Abiotic precipitation, mineral replacement, and fluid flow can produce structures that superficially resemble biological fabrics. The scientific literature contains debated examples from Earth's oldest rocks where proposed microfossils have been reinterpreted as pseudo fossils following more detailed analysis. The interpretive bar for morphology alone, absent chemical and isotopic support, is appropriately high.

Molecular biosignatures are more specific in principle. Life on Earth builds molecules with particular structural regularities that abiotic synthesis tends not to produce in the same proportions: the homochirality of amino acids, the specific chain-length distributions of lipid biomarkers, the functional group patterns that reflect enzymatic biosynthesis rather than random chemical assembly. These signatures can persist for billions of years in favourable conditions, degrading toward simpler fragments and molecular fossils that retain some structural information even as the original compounds break down. The difficulty is that after four billion years of geological processing, the original molecules are unlikely to survive intact. What remains are degradation products whose biological ancestry may be inferable from their distribution patterns but cannot be read off directly from their structures. And on Mars, the additional oxidative processing layer discussed in the preceding chapters means that even these degradation products face ongoing attack. The molecular record at exposed Martian surfaces is not simply ancient and fragmented, it is ancient, fragmented, and under continuing chemical assault.

Isotopic biosignatures exploit a well-documented property of biological metabolism: enzymes preferentially process lighter isotopes of carbon, sulphur, and other elements, because lighter isotopes form slightly weaker bonds and react fractionally faster in enzymatic systems. Microbial carbon fixation consistently produces organic matter depleted in carbon-13 relative to the inorganic carbon source by roughly 20 to 30 parts per thousand, a signal large enough to be measured precisely and distinctive enough to stand out against abiotic background variation in most geological contexts. This isotopic signature can survive even after the organic molecules themselves are destroyed, preserved in secondary carbonates, graphite, or other mineral phases that formed in isotopic equilibrium with the biologically processed carbon. On Mars, measuring carbon isotope ratios with the precision required to detect a biological signal is within reach of existing instrumentation ,Curiosity's Sample Analysis at Mars suite has measured carbon isotope ratios in released gases from heated samples ,but the interpretation requires careful elimination of abiotic fractionation pathways, including photochemical fractionation in the atmosphere and isotopic effects from high-temperature volcanic degassing, that can produce carbon-13 depletions overlapping with the biological range.

Mineralogical biosignatures arise from the way metabolizing organisms interact with their geochemical environment. Iron-reducing bacteria alter ferric iron minerals to ferrous phases as a byproduct of anaerobic respiration, producing distinctive reduced-iron mineral assemblages in their immediate vicinity. Sulphate-reducing bacteria generate sulfide minerals with characteristic textures and sulfur isotope compositions. Microbes can nucleate mineral precipitation on their cell surfaces, producing fabrics that sometimes preserve cell outlines even after the organic material is gone. These metabolic byproducts can be specific enough, when their spatial distribution and isotopic composition are considered together, to provide strong circumstantial evidence for biological involvement. On Earth, these mineralogical biosignatures form the core of the case for life in rocks more than three and a half billion years old, where molecular evidence has been largely lost to metamorphism.

The Cheyava Falls sample, examined by Perseverance in Jezero Crater in 2023 and 2024, illustrates both the promise and the difficulty of this approach on Mars. The sample contains vivianite, an iron phosphate mineral, and greigite, an iron sulfide, arranged in a spatial pattern described as leopard spots ,small reduced-iron zones set within a more oxidized matrix. On Earth, this kind of millimeter-scale redox patterning, with pockets of reduced iron minerals embedded in oxidized rock, is frequently produced by microbial activity: iron-reducing bacteria create localized reducing microenvironments around their cells, generating exactly the kind of patchy reduced-iron mineralogy seen in Cheyava Falls. This is a genuine and interesting observation. It is also not conclusive. Abiotic redox reactions driven by spatial heterogeneity in groundwater chemistry, mineral composition, or fluid pathways can produce similar patterning without any biological involvement. Serpentinization reactions and other water-rock interactions generate strong local reducing conditions capable of precipitating iron sulfides in patterns that can resemble biologically induced features. The community debate over Cheyava Falls has centered precisely on whether the spatial scale, morphology, and mineral assemblage of the leopard spots require biological explanation or whether the abiotic alternatives remain equally plausible given the data available from the rover. That debate has not been resolved, and it is unlikely to be resolved without returning the sample to Earth for the kind of multi-technique analysis that Mars instruments cannot yet provide.

What Cheyava Falls illustrates, beyond the specific question of whether it contains biosignatures, is the ambiguity inherent in interpreting mineralogical evidence from a single rover with limited analytical capability at a site whose preservation history is incompletely characterized. It is a useful test case for what Mars biosignature science actually looks like in practice, as opposed to how it is sometimes described in general terms. The features observed are consistent with biology. They are also consistent with abiotic geochemistry. Strengthening the biological case requires additional independent lines of evidence, and weakening it requires demonstrating that a specific abiotic mechanism can quantitatively reproduce all the observed features in the geological context of that rock. Neither has been accomplished yet. The honest assessment is that Cheyava Falls is an intriguing ambiguity, not a detection.

This pattern ,intriguing ambiguity rather than clear detection ,is likely to characterize Mars biosignature science for some time, and for reasons that go beyond instrument limitations. The ancient rocks that might contain biosignatures have not been sitting undisturbed since they formed. They have been buried, exhumed, fractured, heated, and for however long they have been exposed at the modern surface, chemically processed by an oxidizing atmosphere. Each of these events modifies the original chemical and mineralogical record in ways that can mimic, dilute, or destroy biosignatures. The modern oxidative environment is one layer of this overprinting, but it is the one most directly relevant to what missions can detect today, because it has been operating continuously on whatever the surface currently exposes. The question is not only whether biosignatures formed and survived the deep geological past ,it is whether they have survived the more recent and ongoing chemistry of the present.

Chapter 5: Oxidation Mechanisms and Ratcs

The question of how rapidly oxidation proceeds at the Martian surface, and how deeply it penetrates into exposed materials, is central to assessing preservation. If oxidation under Mars conditions is extremely slow ,negligible on timescales of millions of years ,then the preservation concern raised in previous chapters is more theoretical than practical, and ancient materials exposed at the surface today might retain their original chemistry reasonably well. If oxidation proceeds at rates that are significant over geological exposure times, then the depth and duration of surface exposure become critical variables that exploration strategy must account for. The honest answer, given current data, lies between these extremes and remains genuinely uncertain at the level of detail that would allow confident predictions at specific sites. What laboratory experiments and surface observations do establish is the range of relevant processes, their dependencies on environmental conditions, and the approximate scales over which they operate.

The primary oxidizing agents at the Martian surface are perchlorate and related chlorine-oxygen species, hydrogen peroxide, and superoxide radicals. Perchlorate itself is a relatively stable oxidizer at low temperatures and does not react rapidly with organics under purely thermal conditions at Mars surface temperatures, which range from roughly 150 Kelvin at winter polar regions to around 300 Kelvin at the equatorial surface during peak afternoon heating. However, perchlorate becomes significantly more reactive in the presence of ultraviolet radiation and iron oxides, both of which are abundant at the Martian surface. UV photons at wavelengths below about 300 nanometers ,wavelengths that Mars surface receives in substantially greater flux than Earth's surface does, because the thin Martian atmosphere provides minimal UV shielding ,can photolyze perchlorate and generate reactive chlorine radicals that attack organic molecules far more aggressively than thermally activated perchlorate reactions alone. Iron oxides, the minerals responsible for Mars's characteristic red color, catalyze the breakdown of hydrogen peroxide into hydroxyl radicals, which are among the most reactive oxidizing species in chemistry. The combination of UV radiation, perchlorate, and iron-oxide catalysis creates a surface oxidation environment that is more aggressive in practice than any single component considered in isolation.

Laboratory experiments designed to characterize this combined effect have produced results worth examining carefully. When organic compounds representative of potential Martian biosignatures ,amino acids, carboxylic acids, aromatic hydrocarbons ,are exposed to UV radiation in the presence of perchlorate at Mars-relevant temperatures and pressures, degradation proceeds measurably over experimental timescales of days to weeks. Extrapolating these laboratory rates to Mars surface conditions requires applying Arrhenius temperature corrections, accounting for actual UV flux levels at specific latitudes and seasons, and acknowledging the difference between the continuous UV exposure in a laboratory chamber and the day-night cycling and seasonal variation on Mars. These extrapolations introduce uncertainties that can span an order of magnitude or more in either direction. The resulting estimates suggest that organic molecules at or very near the Martian surface ,within the top few millimeters to centimeters ,face degradation on timescales ranging from thousands to perhaps hundreds of thousands of years under typical equatorial conditions, with slower rates at higher latitudes and in shadowed or subsurface locations. These are geological eyeblinks compared to the age of Noachian or Hesperian sediments, which means that any organics present in ancient rocks exposed today at the Martian equatorial surface have had to survive an enormously long period of oxidative attack since exhumation.

Penetration depth is a critical variable because it determines how far below the surface ancient chemistry is protected from modern oxidative processing. Gas-phase diffusion of reactive species into porous rock or unconsolidated regolith is governed by Fick's law, with penetration scaling roughly as the square root of time at constant diffusivity. Diffusivity in turn depends on pore structure, tortuosity, and the reactivity of the diffusing species with mineral surfaces along the pathway ,a highly reactive species like a hydroxyl radical is consumed rapidly near the surface and never penetrates far, while more stable species like perchlorate can diffuse further before being destroyed. Laboratory measurements of oxidant penetration into Mars regolith simulants suggest active oxidation zones extending centimeters to perhaps tens of centimeters under sustained exposure, with less porous or more reactive lithologies showing shallower penetration than more permeable or less reactive ones. Scaling these laboratory penetration rates to geological exposure times ,hundreds of millions to billions of years in the case of ancient exhumed surfaces ,is less straightforward, because the same process that drives penetration deeper over time also depletes the concentration gradient driving it, creating a self-limiting behavior. Current estimates, combining diffusion models with laboratory kinetics, place the practically significant oxidation depth in most Mars-relevant lithologies at somewhere between a few centimeters and a few meters, with the lower end of this range better supported by direct measurement and the upper end more model-dependent.

Temperature variation across the planet creates a corresponding variation in oxidation rates through the Arrhenius dependence of reaction kinetics on temperature. Regions near the equator, where afternoon surface temperatures can reach 280 to 300 Kelvin in summer, drive chemical reactions substantially faster than polar regions where temperatures rarely exceed 230 Kelvin. This means that the same organic compound, in otherwise identical mineral context, would degrade faster in equatorial terrain than at high latitudes, and that high-latitude or polar sites offer a kinetic preservation advantage that is independent of dust activity. The magnitude of this effect depends on the activation energy of the relevant reactions, which varies by compound, but for typical organic oxidation reactions the rate difference between 300 Kelvin and 230 Kelvin is roughly a factor of ten to one hundred, a meaningful difference when integrated over geological time.

The role of liquid water deserves specific attention because it is easy to default to Earth-centric intuitions about water as a universal facilitator of geochemistry. Under present Mars surface conditions, liquid water is at best transient and highly localized ,thin brine films on mineral surfaces during brief warm humid periods at mid-latitudes, or seasonal near-surface liquid water in perchlorate-rich regolith under narrow temperature and pressure windows. In environments where liquid water is present, even transiently, it can dissolve perchlorate and other oxidants and transport them into mineral pores faster than gas-phase diffusion alone, potentially extending the oxidation zone deeper during wet periods. It also provides a medium for acid-base and redox reactions that proceed negligibly in the dry state. Whether the net effect of episodic liquid water at the modern Mars surface is to accelerate organic degradation or to preserve organics through specific mineral encapsulation reactions ,some minerals, like carbonates and sulfates, can encapsulate organic molecules and protect them during precipitation ,depends on local chemistry and is not the same at all sites. The point is that water's role in Mars surface oxidation chemistry is not simply to accelerate everything, as it would in a terrestrial analogy, but to modulate specific reactions in ways that depend on what the water carries and what mineral phases are present.

Iron mineralogy interacts with oxidation in ways that go beyond simple catalysis. The surface color of Mars ,the rust-red that has prompted its name in every human language that has had one ,comes from nanophase iron oxides coating mineral grains. These coatings, and the more crystalline iron oxide and oxyhydroxide minerals present in larger quantities at depth, participate actively in electron-transfer reactions between oxidants and reductants in the regolith. In a geochemical system containing both oxidized and reduced iron phases alongside organics, the iron can shuttle electrons from organic molecules to atmospheric or dissolved oxidants, effectively serving as a catalytic intermediary for organic oxidation without being net consumed in the process. The efficiency of this catalysis depends on mineral grain size, crystallinity, and surface area;amorphous iron phases with high surface area are more reactive than well-crystallized phases with lower surface area ,and these properties vary across Mars depending on the geological history of the material.

What the combined picture of oxidation mechanisms suggests is a surface environment where the intensity of chemical attack on ancient organisms is real and significant but not instantaneous or spatially uniform. Materials at the very surface, especially at low latitudes, in dusty high-activity regions, and in porous or iron-oxide-rich lithologies, face the most aggressive chemical environment. Materials protected by even a few tens of centimeters of overburden, or located in low-activity regions at high latitudes, face a substantially less aggressive environment. The cumulative oxidative dose a sample has received depends on its full exposure history ,how long it has been at the surface, where it has been, what it is made of, and how deep within its host rock it sits. None of these variables is directly readable from current orbital data alone. Estimating them at any candidate landing site requires combining surface age estimates from crater counting, dust activity characterization from imaging surveys, rock type identification from spectroscopy and context imaging, and some assumptions about exhumation history. This is a solvable problem, but it requires treating preservation assessment as a quantitative discipline rather than a qualitative judgment that ancient rocks at a watery site should be good candidates.

Chapter 6: Preservation Patterns Across Mars

The previous chapters have established the mechanisms by which modern oxidative chemistry operates at the Martian surface and the factors that control its rate and penetration depth. What this chapter addresses is how those mechanisms combine, at the scale of the whole planet, to create a spatial pattern of preservation quality that is structured, physically interpretable, and consequential for exploration planning. The preservation gradient concept ,that ancient biosignature material survives better in some locations than others, for reasons related to modern atmospheric processing rather than just ancient geological history ,needs to be developed carefully here, because it is the framework's core testable prediction and because being imprecise about what it does and does not claim is a straightforward way to undermine a genuinely useful idea.

The starting point is that preservation quality at any given surface location on Mars is determined by the interaction of at least four variables: the intensity of dust activity over geological time, the duration of surface exposure since last exhumation, the physical and mineralogical properties of the host rock, and the depth below the surface at which the sample of interest sits. These variables are not independent ,rock permeability, for example, links rock type to oxidant penetration depth ,but separating them conceptually is useful for understanding which ones can be characterized remotely and which require surface access to constrain.

Dust activity is the variable most directly tied to the electrochemical pathway that produces and distributes oxidants. The Context Camera survey data discussed in Chapter 3 provides the most comprehensive current characterization of how dust devil activity is distributed across Mars, and it shows clear geographic structure. The

Amazonian plains in the northern hemisphere, large sections of the Tharsis plateau flanks, and certain mid-latitude regions show substantially elevated dust devil detection densities compared to heavily cratered southern highland terrain or high-latitude zones where seasonal temperature extremes limit convective boundary layer development. This is not a subtle difference in the data ,some regions show vortex detection densities an order of magnitude higher than others. If oxidant production rates scale roughly with vortex frequency and intensity, as the triboelectric mechanism predicts, then the cumulative oxidative dose delivered to surface materials over millions of years differs substantially between these regions. The relationship between detection density and actual oxidant production rate involves assumptions that have not been directly calibrated ,it requires knowing how much perchlorate each vortex produces per unit time, how efficiently produced perchlorate is deposited locally versus transported downwind, and how the production rate scales with vortex size and intensity. These are open quantitative questions. The qualitative conclusion that high-activity regions deliver more cumulative oxidative processing than low-activity regions is robust; the specific magnitude of the difference awaits direct measurement.

Exposure duration is the second major variable, and it interacts with dust activity in a straightforward way: a highly active region that has been exposed for only a few million years may have delivered less total oxidative dose to surface materials than a moderately active region exposed for a billion years. Surface age can be estimated from crater size-frequency distributions ,younger surfaces show fewer impact craters per unit area, because they have accumulated fewer craters since their formation or last resurfacing. This method has well-understood uncertainties at different age ranges, performing best for surfaces between roughly a hundred million and a few billion years old, and provides a basis for estimating relative exposure durations across Mars. Combining activity maps with surface age maps gives a first-order proxy for cumulative oxidative dose, even before any direct chemical measurements are made at the candidate sites.

Rock type and permeability create a third axis of variation. A fine-grained basalt with minimal porosity presents a different oxidation target than a permeable sandstone or a fractured mudstone with abundant clay-filled grain boundaries. In the basalt, reactive species reaching the surface are consumed in a thin rind and penetrate slowly inward; in the sandstone, pore connectivity allows both gas-phase diffusion and episodic brine transport to reach centimeters to decimeters of depth. The orbital spectroscopy datasets discussed in Chapter 3 provide mineralogical information relevant to rock type identification, though at spatial resolutions that average over areas of hundreds of square meters to square kilometers and cannot resolve meter-scale lithological variation. Identifying the dominant rock type at a candidate site is possible from orbit; characterizing the detailed pore structure and permeability that govern oxidant penetration requires surface access.

The fourth variable ,sample depth ,is in some ways the most operationally significant, because it is the one that mission design can directly address. Even at a site that scores poorly on the first three variables ,high dust activity, long surface exposure, permeable host rock ,the material at sufficient depth has been protected from surface oxidative chemistry by the overlying rock column. How much depth is sufficient depends on the specific combination of the first three variables: in the worst-case scenario of highly permeable rock with long exposure and intense dust activity, significant penetration depths of a meter or more might be needed to access pristine material; in more favorable combinations, tens of centimeters might be adequate. Current Mars surface missions lack deep drilling capability. The InSight lander's mole ,a self-penetrating probe designed to reach five meters depth to measure heat flow ,failed to achieve its design depth due to regolith properties unforeseen from orbital data. Perseverance and Curiosity drill to a few centimeters depth to collect sample cores. Accessing the depths where the preservation framework predicts protected material exists requires engineering capability that no currently operating Mars mission possesses.

Several geological settings stand out as candidates for enhanced preservation based on these considerations. Impact craters large enough to excavate material from depths of tens of meters or more during formation expose material that was protected by overlying rock for most of its history, but this material is then exposed at the surface in the ejecta blanket and crater wall outcrops. The preservation advantage of deep excavation is offset by the subsequent exposure, and the time since the impact determines how much post-excavation oxidative processing has occurred. Fresh craters ,those formed in the last few tens of millions of years ,offer the best combination of deep excavation and short post-excavation exposure time. Targeting fresh craters, particularly those that have been excavated into ancient sedimentary sequences at depth, provides one strategy for accessing protected material without requiring deep drilling.

Locations where geologically recent processes have buried ancient terrains under younger, less permeable material offer another category of enhanced preservation. Volcanic resurfacing events that covered ancient sediments with lava flows create a cap that excludes atmospheric contact and limits oxidant penetration from above. Where erosion has subsequently cut through the volcanic cap and exposed the ancient sediments at cliff faces or in crater walls, the material may have been protected for most of the intervening time. The key question is how much time has passed since the protective cap was breached by erosion and exposure began. High-resolution surface age estimation in the immediate vicinity of such exposures, combined with the oxidation depth estimates discussed in Chapter 5, can constrain whether subsurface material below the current erosion surface remains in the protected zone.

High-latitude regions offer a kinetic preservation advantage through lower average temperatures, as discussed in Chapter 5. The tradeoff is that polar and near-polar terrains may not be the locations where ancient aqueous habitability was greatest ,warmer equatorial environments with access to liquid water were probably more biologically productive in the Noachian and early Hesperian, if life existed at all. A low-latitude site with excellent ancient geology but high modern processing rates, versus a high-latitude site with more modest ancient geology but better preservation kinetics, presents a genuine strategic tradeoff that cannot be resolved without better data on both the ancient biological potential and the modern preservation conditions at specific candidate sites. The framework does not dictate a single answer ,it insists that both variables be evaluated rather than only one.

Mapping preservation potential across Mars in a quantitative way is technically within reach given existing and near-term datasets. Dust devil activity maps from Context Camera analysis provide the primary input for oxidant production estimates. Surface age maps from crater counting provide exposure duration estimates. Mineralogical maps from the Compact Reconnaissance Imaging Spectrometer provide rock type proxies. Digital elevation models provide topographic context that modulates local dust activity and deposition patterns. Combining these datasets into a preservation probability map ,with explicit uncertainty estimates at each location reflecting the gaps in the underlying data ,would provide a more rational basis for landing site selection than the current approach, which treats preservation quality as a secondary consideration relative to ancient geological interest.

The ultimate test of whether the preservation gradient framework is correct is not a map made from orbital data. It is a comparison of actual measured organic content and biosignature quality at multiple sites specifically chosen to span the range of predicted preservation conditions. If sites predicted to have good preservation ,low dust activity, short exposure, low-permeability host rock, or adequate sample depth ,consistently show higher organic abundance and better-preserved biosignature indicators than sites predicted to have poor preservation, the framework is validated in the way that matters for science. If the correlation is absent, the framework needs revision. Building this comparative dataset is the mission design challenge that the preservation gradient hypothesis motivates most directly, and it is one that can only be met by coordinating landing site choices across multiple missions with this specific comparison in mind, rather than selecting each mission's target independently on its own geological merits.

Chapter 7: Cold Chemistry and Subsurface Brines

Perchlorate's presence in Martian regolith matters for more than oxidation chemistry. Perchlorate salts are powerfully hygroscopic and dramatically depress the freezing point of water, enabling liquid brines at temperatures where pure water would be locked solid. This property connects the dust-driven electrochemistry discussed in earlier chapters to a separate question: whether liquid water exists anywhere on present-day Mars, and if so, what its chemistry looks like and what it means for both habitability and subsurface preservation. The answers are genuinely uncertain, and the uncertainty runs in both directions. The conditions that might allow liquid brines to persist are real and physically grounded, but the evidence that they actually do persist at accessible depths is weaker than it is sometimes presented.

The thermodynamic case for perchlorate brines begins with laboratory measurements of freezing point depression. Calcium perchlorate solutions remain liquid to approximately minus 74 degrees Celsius at eutectic composition. Magnesium perchlorate brines stay liquid to around minus 57 degrees Celsius. Sodium perchlorate brines have a eutectic near minus 36 degrees Celsius, substantially warmer than the calcium and magnesium forms. These are not exotic laboratory curiosities ,they are straightforward consequences of colligative properties, as well-established as any physical chemistry. If perchlorate salts at these concentrations are present in Martian regolith, and if water is available to dissolve them, the resulting brines can remain liquid at temperatures routinely reached in the Martian shallow subsurface during summer at mid-latitudes.

The source of water for such brines would be atmospheric. Mars carries trace amounts of water vapor ,global column abundances averaging around 10 precipitable micrometers, varying seasonally and with latitude ,and hygroscopic minerals can scavenge this moisture from the air through deliquescence, dissolving in the thin water film they pull from the atmosphere. Perchlorate is among the most effective deliquescent salts known, capable of absorbing water from the air at relative humidity levels well below what would cause condensation on a non-hygroscopic surface. Whether atmospheric moisture is sufficient to produce brines rather than just hydrated salt phases at Mars-relevant humidity levels is a quantitative question whose answer depends on local perchlorate concentration, temperature, and the daily and seasonal humidity cycle at the site in question. Modeling studies suggest that in certain mid-latitude regions during summer, conditions are intermittently favorable for brine formation at depths of a few centimeters to tens of centimeters, but these models carry significant uncertainty because the relevant parameters ,perchlorate concentration at depth, pore-scale humidity, thermal gradient through the regolith ,are not well constrained by available measurements.

The recurring slope lineae controversy illustrates how the brine question sits between plausible hypothesis and demonstrated fact. These features ,dark streaks that appear on steep slopes during warm seasons and fade or disappear during cold seasons at sites across Mars ,were first interpreted as possible evidence of transient brine flows when detailed spectroscopic analysis suggested hydrated perchlorate salt signatures associated with some of them. The seasonal timing, the spectroscopic hint of hydration, and the association with steep sun-facing slopes in mid-latitude and equatorial regions all fit a narrative of brines forming during warm periods and evaporating or freezing as temperatures drop. That interpretation attracted substantial attention and generated genuine excitement about present-day liquid water on Mars.

Subsequent analysis has substantially complicated the picture. Work published by Dundas and colleagues in 2017 and in follow-on studies showed that the geometric properties of recurring slope lineae

,their slope angles, lengths, and behavior at cliff edges ,are consistent with dry granular flows, in which loose sand or dust cascades downhill without any liquid involvement. Many lineae occur on slopes steep enough that dry material would be expected to flow given the right triggering conditions, such as frost sublimation or seismic disturbance. The spectroscopic hydration signatures, initially interpreted as perchlorate brines, were subsequently reanalyzed and found in several cases to be consistent with ferric minerals that produce similar spectral features without requiring liquid water. Not all recurring slope lineae have been analyzed in equivalent detail, and the community has not declared the brine hypothesis definitively dead, but the current weight of evidence favors predominantly dry mechanisms for most of the observed features. Liquid water involvement cannot be ruled out in all cases, but it is no longer the leading interpretation for the class of features as a whole. A book that presents the brine hypothesis and the dry-flow alternative as equally supported is accurately representing the current state of the science; a book that presents brines as the established explanation for recurring slope lineae is not.

Setting aside recurring slope lineae specifically, the question of whether perchlorate brines exist at centimeter-to-meter depths in the Martian regolith remains open and interesting. The thermodynamic conditions for brine stability are met intermittently in model calculations at mid-latitude sites during summer. The deeper question is whether the system actually reaches thermodynamic equilibrium ,whether enough water is available at the right places and times to dissolve the perchlorate and form a liquid phase, rather than simply maintaining hydrated but still-solid salt phases. This is a question that ground truth measurements could answer, and none of the instruments deployed on Mars so far has been well-suited to answering it. Ground-penetrating radar at the resolution needed to detect centimeter-scale liquid layers in the shallow subsurface has not been flown on a landed mission. Direct sampling at depths of tens of centimeters with chemical analysis capable of distinguishing liquid from solid phases is beyond current rover drilling capability. The InSight lander's heat flow probe ,which would have provided subsurface thermal data relevant to brine stability modeling ,failed to reach its design depth due to unexpectedly cohesive regolith. The data gap is real, and it limits how confidently the brine hypothesis can be either confirmed or dismissed.

For the preservation framework developed in earlier chapters, whether subsurface brines actually form has a specific and consequential implication. The oxidation penetration depth estimates discussed in Chapter 5 were based primarily on gas-phase diffusion of oxidizing species into porous rock. If liquid brines exist at depth, even episodically, they represent a far more efficient transport mechanism for dissolved oxidants. A perchlorate brine layer at thirty centimeters depth does not just sit there passively ,it is itself a concentrated oxidant solution in contact with whatever rock and organic material surrounds it. The presence of brines would extend the chemically active zone substantially deeper than gas diffusion alone would suggest, in ways that depend sensitively on brine depth, seasonal duration, and the oxidant concentration in the brine. This creates a scenario where the oxidation depth is not a single number with reasonable uncertainty bounds but a variable that depends on whether liquid phases form at all, and when, and at what composition. The honest position is that this complication exists and matters, but its actual magnitude is not currently constrained by direct measurement.

From a resource standpoint, perchlorate brines at accessible depths would represent a combined source of water and oxidant that is in principle extractable. The energy accounting is worth being clear about. Extracting water from a perchlorate brine requires either evaporating the water selectively ,energy-intensive but technically straightforward ,or running the brine through a separation process such as reverse osmosis or electrodialysis, both of which require power and specialized membranes or electrodes. The water content of a concentrated perchlorate brine is substantially lower per kilogram of processed material than the water content of a water-ice deposit at comparable depth, meaning brines are a less mass-efficient water source than ice if ice is available at the same site. The perchlorate recovered in the process has genuine value as a feedstock for oxygen production via thermal decomposition, which partially offsets the processing energy cost. Whether brine extraction competes favorably against other water acquisition strategies depends on site-specific resource availability and mission energy budget ,at high-latitude sites where water ice is accessible near the surface, brine extraction is probably not the preferred option; at lower-latitude sites where surface ice is absent and atmospheric water extraction is limited by low humidity, brines at accessible depth might be competitive if they actually exist there.

The planetary protection dimension of the brine question deserves straightforward treatment. COSPAR planetary protection policy categorizes Mars as requiring special precautions against forward contamination ,the inadvertent delivery of viable Earth organisms to Mars that could compromise the search for indigenous Martian life. Environments where liquid water might exist are categorized as special regions requiring the highest level of contamination control. If perchlorate brines at centimeter-to-meter depths are common across mid-latitude Mars, a large fraction of potential landing sites could intersect special region conditions, substantially complicating mission planning and potentially ruling out certain surface activities. If brines are rare or absent at accessible depths, the special region concern is correspondingly reduced. Resolving the brine question is therefore not only scientifically interesting but operationally important for planning human Mars missions, where crew activities inevitably involve contact with the near-surface regolith.

What the current evidence supports is this: the physical and chemical conditions for perchlorate brine formation at shallow

Martian subsurface depths are intermittently plausible at certain latitudes and seasons based on thermodynamic modeling and measured perchlorate abundances. Direct evidence that liquid brines actually form and persist at those depths is lacking, and the surface features previously most cited as brine evidence ,recurring slope lineae ,are now better explained by dry mechanisms in most cases. The question is open and measurable, and resolving it should be a specific objective of future subsurface-access missions rather than an assumption carried uncritically in either direction.

Chapter 8: Process-Driven Site Selection

Every Mars landing mission to date has selected its site primarily on the basis of ancient geological interest. The logic is straightforward and defensible: if the goal is to find evidence of past life, go where conditions were favorable for life in the past, which means going where water was present and where habitability indicators ,clay minerals, carbonates, sulfates, delta deposits ,are preserved in the geological record. This reasoning has guided missions from Viking through Perseverance, and it has produced genuine scientific returns. Ancient lake beds have been confirmed, delta stratigraphy has been characterized, and organic molecules have been detected in Noachian and Hesperian sediments that formed when Mars was a different planet than it is today.

The argument developed through the preceding chapters does not dispute the value of ancient geological interest as a site selection criterion. It argues that it is insufficient on its own, because it treats preservation quality as uniform across candidate sites when there is physical reason to expect it is not. A site with exceptional ancient geological credentials but a long history of intense dust activity, long surface exposure, and permeable host rock has been subject to far more cumulative oxidative processing than a site with more modest ancient features but low dust activity, short exposure, and tight lithology. The probability of finding intact biosignatures, if they ever existed, is different at these two sites in a way that current site selection practice does not account for.

Translating the preservation gradient framework into an operational site selection approach requires being honest about which components of the framework can currently be quantified from orbital data and which require assumptions that introduce genuine uncertainty. Dust activity characterization from Context Camera surveys is the most mature component. Multi-year surveys of dust devil track density and active vortex detections provide spatially resolved maps of relative activity intensity across large portions of the planet. Converting detection density to an absolute oxidant production rate requires knowing how much perchlorate each vortex produces per unit time as a function of size and intensity ,a calibration that laboratory data can inform but that has not been validated in the field. The orbital data can reliably rank candidate sites by relative dust activity; absolute processing rates remain estimates with order-of-magnitude uncertainty.

Surface age estimation from crater size-frequency distributions adds a second layer of information. Younger surfaces have accumulated fewer impact craters per unit area since their last resurfacing event, and crater counting statistics can constrain model ages for geological units visible from orbit. The method has known uncertainties ,the crater production rate at Mars is calibrated through a chain of assumptions connecting lunar chronology to Mars via meteorite ages, and small crater populations are affected by secondary cratering and erasure that introduces additional scatter ,but it provides relative age constraints that are useful for ranking candidate sites by estimated exposure duration. A site where ancient sediments were covered by a younger lava flow and only recently re-exposed by erosion has had less time accumulating oxidative dose than a site where equivalent sediments have been at the surface since the Hesperian. These distinctions are readable, with appropriate uncertainty, from orbital data.

Rock type and permeability are harder to constrain remotely. The Compact Reconnaissance Imaging Spectrometer provides mineralogical information that correlates broadly with lithology ,clay-bearing units tend to be fine-grained and relatively low-permeability, sulfate-bearing units are often evaporitic and potentially quite permeable, and so on ,but the spectroscopic classification does not translate directly into quantitative permeability values relevant for oxidant diffusion modeling. This is a limitation that accepting and accounting for uncertainty estimates, rather than papering over, is the honest approach. Site selection informed by the preservation framework should acknowledge that rock permeability estimates from orbit carry large uncertainty and that surface measurements at candidate sites would substantially improve preservation quality predictions before committing to a final target.

The interaction between these variables and mission capability creates a set of explicit tradeoffs that site selection committees should be able to reason about quantitatively rather than qualitatively. A mission with no subsurface drilling capability is entirely dependent on surface and near-surface samples, meaning it is most exposed to the effects of surface oxidation and most benefits from sites with low cumulative oxidative dose. For such a mission, a moderate-quality ancient site with low dust activity and short exposure history may return better biosignature science than a spectacular ancient site that has been heavily processed. A mission with meaningful drilling capability ,access to depths of a meter or more ,can offset poor surface preservation by accessing protected material below the oxidation zone, which reduces the priority of minimizing dust activity in site selection and shifts emphasis back toward ancient geological quality. A sample return mission capable of caching samples from multiple locations at a single site could deliberately sample the depth profile at a preserved ancient target, testing how preservation quality varies with depth in a way that directly validates the oxidation penetration estimates on which the framework depends.

There is also a category of site that the preservation framework identifies as particularly promising but that conventional ancient-geology-focused selection tends to undervalue: locations where ancient sediments have been recently exhumed from beneath protective younger cover. Erosional windows into ancient stratigraphy ,places where a younger lava flow, volcanic ash deposit, or impact ejecta blanket has been cut through by erosion to expose the ancient unit beneath ,present material that was shielded from atmospheric chemistry for most of its history and is only now being exposed. The preservation advantage of this recent exhumation is time-limited; once exposed, the material begins accumulating oxidative dose at whatever rate is appropriate for that location. But a freshly exhumed surface that has been exposed for only tens of millions of years rather than billions represents a substantially lower cumulative dose, even at a moderately active site, than a surface that has been exposed since the Hesperian. Fresh impact craters that have punched through younger cover into ancient stratigraphy represent natural drilling events that bring deep material to the surface in the ejecta blanket and crater wall outcrops, with exposure ages constrained by the crater's own age.

Incorporating these considerations into operational site selection does not require a new institutional framework. It requires adding data layers to the geographic information systems already used in site selection workshops: a dust activity index map from Context Camera analysis, a surface age map from crater counting and geological mapping, a lithological proxy map from spectroscopy and thermal inertia, and an estimated cumulative oxidative dose map synthesized from the first two inputs. These layers would allow site selection teams to visualize the tradeoff between ancient geological interest and predicted preservation quality explicitly, rather than leaving it as an intuitive and unquantified judgment. The scoring functions used to rank candidate sites would need to include preservation quality as a weighted factor alongside trafficability, landing safety, and scientific diversity ,with the weight assigned to preservation quality being a mission-design choice that reflects the specific science objectives of the mission.

The framework predicts a specific and testable outcome from missions designed with this in mind: sites ranked as having better preservation should return samples with higher organic abundance and better-preserved biosignature indicators than sites ranked as having poorer preservation, controlling for ancient geological context. If this correlation is found, it validates the approach and strengthens the case for using it in subsequent mission planning. If no correlation is found, the framework needs revision ,perhaps the magnitude of oxidative processing differences between sites is smaller than modeled, or perhaps other preservation mechanisms dominate. Either outcome advances the science. What does not advance the science is continuing to select sites without accounting for modern preservation conditions and then finding variable biosignature quality across missions without a framework for understanding why.

Chapter 9: Electrostatic Extraction Systems

The standard conceptual model for resource extraction on Mars borrows heavily from terrestrial mining: mechanical excavation of bulk regolith, crushing and sorting to concentrate target materials, and thermal or chemical processing to extract specific compounds. This approach has the advantage of extensive terrestrial precedent and well-understood engineering principles. It has the disadvantage that the equipment required ,excavators, conveyors, crushers, reactors ,is heavy, mechanically complex, and energy-intensive. Delivering and powering that infrastructure in early Mars missions, when mass budgets are severely constrained and power systems are limited, is a genuine challenge. The question worth asking is whether Mars offers alternatives that leverage what the planet already does rather than importing complete industrial systems.

The triboelectric charging that makes dust devils electrochemically interesting also makes them, in principle, sorting machines. When particles of different sizes and compositions collide inside a vortex, they separate charge based on differences in electron affinity, and the resulting charge-to-mass ratio differences cause the electric fields within the vortex to segregate them. Smaller, finer particles tend toward one polarity; larger, coarser particles toward the other. Over the lifetime of a dust devil, this sorting process redistributes material by both size and, to some degree, composition. The planet does this across millions of vortices every active season, without any human infrastructure. The question for resource extraction is whether systems can be designed to intercept, redirect, or amplify this natural sorting to accumulate materials of interest in concentrations higher than their background distribution in bulk regolith.

It is important to be clear about the technology readiness level of this concept. Electrostatic separation of minerals is a real and industrially practiced technique on Earth, used in processing beach sand mineral assemblages, separating conductor from non-conductor mineral phases in heavy mineral concentrates, and cleaning coal and other bulk materials. The physics is well-established. Applying it to Mars involves translating a terrestrial industrial process into a low-pressure, low-gravity, extreme-temperature environment with limited power and maintenance capability. That translation has not been demonstrated at Mars-relevant scale in the field, and laboratory demonstrations under Mars-analog conditions ,while they exist and validate the basic physical mechanism ,do not constitute proof that operational systems with useful throughput are achievable. These concepts currently sit at roughly Technology Readiness Level 2 to 3: the working principle is demonstrated in laboratory experiments, but no integrated system has been tested under Mars-analog conditions outside the laboratory, let alone in actual Mars surface conditions. The value of the idea is real, but its maturity as an engineering solution is low.

With that caveat explicit, the concept architecture is worth developing because it describes what would need to be demonstrated for electrostatic extraction to become operationally useful. Passive collectors ,mesh or plate structures positioned to intercept wind-transported dust ,exploit the fact that Martian dust particles arrive pre-charged from atmospheric triboelectric interactions. A grounded conductor or a surface maintained at controlled potential relative to ground presents an electrostatic target that preferentially attracts particles of matching charge polarity. The simplest version requires no power during collection: the structure is deployed, the wind does the work, and charged dust accumulates on collection surfaces. Laboratory experiments with Martian soil simulants in low-pressure chambers confirm that mesh collectors with appropriate geometry accumulate fine dust under wind-analog conditions, with collection efficiency dependent on wind speed, particle size distribution, and mesh spacing. The collection rates measured in these experiments are real but modest, and extrapolating them to useful material throughput at Mars surface conditions requires assumptions about wind regimes and dust flux that carry significant uncertainty.

A more capable active system would apply controlled electric fields across collection electrodes, attracting charged particles of target polarity while repelling others. By switching field polarity or adjusting field geometry, such a system could preferentially accumulate different particle populations in sequence ,for instance, first collecting the fine, negatively charged fraction enriched in perchlorate-bearing particles, then switching to accumulate the coarser, positively charged fraction enriched in silicate minerals. The selectivity achievable depends on how cleanly different mineral and particle-size populations separate by charge polarity in the natural triboelectric process, which varies with local dust composition and atmospheric conditions. This variation is a real source of uncertainty in how well active electrostatic separation can actually fractionate Martian dust into compositionally distinct streams. Power requirements for an active system operating at meaningful throughput are not trivial ,maintaining field potentials across meter-scale electrode arrays in a dusty, ionizing environment involves power costs and insulation challenges that need hardware-level validation before confident estimates are possible.

Perchlorate is a specific target that motivates much of this discussion because of its dual value as an oxygen feedstock and as a water-concentrating agent in hygroscopic brine formation. If natural dust transport and deposition processes preferentially concentrate perchlorate-bearing fine particles in specific depositional environments ,topographic lows, lee sides of ridges, crater floors where settled dust accumulates ,then those environments represent natural pre-concentration targets analogous to ore deposits in terrestrial prospecting. The concentrations achievable through natural deposition are unlikely to be dramatic multiples above the background half-percent-by-weight measured at Phoenix and Curiosity, but even modest enrichments reduce the volume of material requiring processing to extract a unit of perchlorate. Identifying such natural concentration zones requires the same orbital thermal inertia and dust accumulation rate datasets that inform preservation gradient mapping ,another instance where science and resource assessment can share a common observational base.

Thermal decomposition of perchlorate to produce oxygen is the extraction step that follows collection and concentration, and it is the most technically mature element of the chain. Heating calcium or magnesium perchlorate above roughly 400 degrees Celsius produces oxygen gas and the corresponding metal chloride residue. This reaction is well-characterized and has been demonstrated in the laboratory under Mars-relevant conditions. The oxygen yield per kilogram of calcium perchlorate is approximately 0.26 kilograms, meaning that processing a kilogram of regolith at half-percent perchlorate content yields roughly 1.3 grams of oxygen ,a modest yield per kilogram of processed material, but material that costs nothing to mine if the dust is collected from the atmosphere rather than excavated from the ground. Scaled to the oxygen requirements of a small crew ,roughly one kilogram per person per day for breathing, more for oxidizer in propellant production ,the numbers suggest electrostatic dust collection would need to operate continuously at substantial collection rates to make a meaningful contribution. This is an honest scaling exercise that the concept needs to survive before it earns serious mission planning consideration.

The most defensible near-term role for electrostatic dust collection is probably not as a primary resource extraction system but as an adjunct to infrastructure that is deployed for other purposes anyway. Solar panels on Mars accumulate dust and require cleaning to maintain power output. A panel cleaning system that captures removed dust rather than simply dispersing it recovers material that has been naturally sorted and partially processed by atmospheric chemistry. Habitat surfaces, antenna structures, and other large-area deployments present collection surfaces to the wind at no additional mass cost. The material accumulated passively from these surfaces over a mission's operational lifetime might yield meaningful quantities of perchlorate and fine mineral fractions without requiring dedicated collection infrastructure. This is a more conservative claim than presenting electrostatic extraction as a major resource production pathway, and it has the advantage of being testable on near-term missions without requiring a dedicated demonstration mission.

The broader principle that Mars resource extraction should look for ways to interface with what the planet naturally does, rather than importing complete terrestrial industrial solutions, is sound regardless of whether electrostatic collection specifically achieves useful scale. Atmospheric electrochemistry processes material continuously. Wind sorts and transports that material. Gravity deposits it. Any resource extraction approach that can position itself downstream of these natural processes ,intercepting what the planet has already sorted, rather than starting from bulk unprocessed regolith ,reduces the mass and energy cost of initial processing steps. The specific efficiency gains depend on the material and the process, and some will prove more practically useful than others. Electrostatic dust collection is one instance of this broader design philosophy, and it merits continued laboratory development and small-scale flight demonstration before conclusions about its operational utility are drawn in either direction.

Chapter 10: Perchlorate as Process Marker

Perchlorate tends to appear in Mars discussions in one of two roles: as a hazard to be managed or as a resource to be exploited. The hazard framing is straightforward ,perchlorate is toxic to mammals at low concentrations, interfering with iodine uptake by the thyroid gland, and any crew living and working in perchlorate-rich regolith will need stringent protocols for dust control, habitat filtration, and water purification. The resource framing is equally straightforward: heating perchlorate above roughly 400 degrees Celsius releases oxygen gas, making it a potential feedstock for life support and propellant production at a settlement that cannot afford to import consumables from Earth. Both perspectives are legitimate engineering concerns that deserve serious treatment in mission planning.

What both perspectives miss is a third use for perchlorate that is scientifically distinct from either: its value as a tracer for the spatial and temporal pattern of modern atmospheric chemistry. If the triboelectric production mechanism discussed in Chapter 2 is the dominant pathway by which perchlorate accumulates in Martian regolith ,a conclusion supported by laboratory efficiency comparisons but not yet validated by spatially resolved field measurements ,then the distribution of perchlorate across Mars encodes information about where and how intensely the atmosphere has been doing electrochemical work on the surface. A measured perchlorate concentration is not just a number indicating oxidant abundance. It is, under this interpretation, a record of integrated dust activity at that location over the exposure history of the sampled material.

The existing measurements do not yet allow this interpretation to be tested rigorously. Phoenix detected perchlorate at roughly 0.4 to 0.6 percent by weight in polar soil samples at its northern landing site. Curiosity measured comparable concentrations in mudstone and aeolian deposits in Gale Crater at equatorial latitudes. These two sites are separated by thousands of kilometers, differ in latitude, geological setting, and surface age, and sit in regions that orbital dust activity surveys characterize as moderately active ,neither among the most intense dust devil hotspots nor among the quietest terrains on the planet. The fact that they return similar perchlorate concentrations is consistent with several different interpretations. It could indicate that global atmospheric mixing over geological time has homogenized perchlorate concentrations despite spatial variation in local production rates ,a plausible outcome if long-range dust transport is efficient enough to redistribute perchlorate-bearing particles across the planet faster than local production and destruction rates create gradients. Alternatively, it could indicate that local production rates at these two moderate-activity sites happen to be similar. Or it could reflect measurement sampling that is simply too sparse and too geographically clustered near moderate-activity sites to detect the full range of concentrations that actually exists across Mars.

Distinguishing between these possibilities requires measurements that have not yet been made. A dataset containing perchlorate concentrations from sites spanning the full documented range of dust activity ,from the high-activity Amazonian plains and low-thermal-inertia terrains where Context Camera surveys find the densest vortex populations, to the relatively quiescent ancient highland terrains where detections are sparse ,would allow a direct test of whether concentration correlates with activity as the triboelectric production hypothesis predicts. If concentrations at high-activity sites run measurably higher than at low-activity sites, controlling for geological context and sample depth, the tracer interpretation gains strong support. If concentrations prove indistinguishable across sites with very different activity histories, it suggests either that global mixing overwhelms local production differences or that the dominant production mechanism is not locally variable in the way triboelectric discharge would predict.

The depth profile of perchlorate concentration at a single site would add complementary information. If perchlorate is produced at the surface through atmospheric chemistry and then transported downward by diffusion, wicking through brine films, or physical incorporation during periodic dust deposition, the concentration profile should decrease with depth ,highest at or near the surface where production and deposition are strongest, declining toward background levels at depth. If perchlorate is instead distributed relatively uniformly with depth, it might indicate that ancient aqueous processes redistributed it throughout the near-surface regolith before the present arid conditions established themselves, or that downward transport is efficient enough to maintain near-uniform concentrations. A profile that shows elevated concentrations at some specific depth interval rather than monotonically decreasing from the surface might indicate a paleosurface ,a former ground surface now buried ,where perchlorate accumulated during a period of active atmospheric chemistry before subsequent deposition covered it. Each of these profiles tells a different story about the history of perchlorate at that location, and none of them can be distinguished without sampling at multiple depths within a single borehole. Current Mars mission drilling capability, limited to a few centimeters, is insufficient for this kind of profiling.

Chlorate concentration measured alongside perchlorate adds another diagnostic dimension. Chlorate is the penultimate oxidation product in the triboelectric pathway ,the step immediately before perchlorate in the chloride-to-perchlorate oxidation sequence. If active production of oxidized chlorine species is ongoing at a site, chlorate should be present alongside perchlorate because some fraction of the material in the pathway has not yet been fully oxidized to perchlorate. If perchlorate is a static legacy from ancient production with no ongoing replenishment, chlorate concentrations should be relatively low because the reaction sequence has had time to run to completion. Measuring chlorate-to-perchlorate ratios at multiple sites could therefore indicate whether local atmospheric chemistry is actively producing oxidized chlorine at present or whether the perchlorate inventory is largely inherited from past activity. Sample Analysis at Mars on Curiosity has the analytical capability to detect both species in heated samples, though the specific measurements required for this comparison have not been systematically targeted.

Isotopic composition of perchlorate, if measurable, would provide the most direct diagnostic for production mechanism. Perchlorate produced through triboelectric plasma discharge incorporates oxygen atoms from the atmospheric CO_2 and from mineral surfaces in the plasma region, while perchlorate produced through UV photochemistry incorporates oxygen through different reaction pathways. These different oxygen sources have different isotopic compositions, and the fractionation associated with each production mechanism is likely distinct in ways that an isotopic measurement could resolve. Chlorine isotope ratios might similarly vary between production pathways. Measuring these isotopic signatures requires analytical precision well beyond what current Mars surface instruments can achieve, but returned samples analyzed in terrestrial laboratories could provide this data. The Mars Sample Return program, if it succeeds in returning material from multiple geological contexts, could yield the isotopic dataset needed to fingerprint perchlorate production mechanisms ,but only if the sample caching strategy is designed with this question in mind and includes samples from sites with contrasting dust activity histories rather than concentrating all cached material within a single geological setting.

The practical implication of treating perchlorate as a process marker rather than simply a hazard or resource is that it changes what information is extracted from perchlorate measurements and how that information informs mission planning. A perchlorate measurement at a candidate landing site, combined with knowledge of that site's dust activity history and surface age, becomes an estimate of the cumulative oxidative dose that surface materials have received ,relevant both to resource assessment, by indicating how perchlorate-rich extractable material is likely to be, and to preservation assessment, by indicating how aggressively oxidative chemistry has been attacking any organic material present. Getting both pieces of information from a single measurement type makes perchlorate mapping a high-value objective for orbital and landed assets regardless of whether the mission's primary focus is astrobiology, resource prospecting, or human mission planning. The chemistry does not respect those disciplinary boundaries, and neither should the measurement strategy.

One additional dimension of the chlorine chemistry on Mars deserves acknowledgment here, because it affects how perchlorate concentration should be interpreted as a process marker. Recent laboratory and modelling work makes clear that the chlorine cycle on Mars is not one-directional. Galactic cosmic rays penetrate the upper metres of the regolith and drive radiolytic reactions that decompose perchlorate back into lower-oxidation-state species ,chlorate, hypochlorite, chlorine dioxide ,and molecular oxygen. UV radiation at the surface drives both formation and destruction of perchlorate depending on the specific mineral substrate and chloride concentration involved. The practical consequence is that the surface concentration measured by Phoenix and Curiosity represents not the total perchlorate produced by triboelectric chemistry over the exposure history of that material, but the net result of production and destruction running simultaneously. The fact that both sites return roughly half a percent by weight, across different geological settings and different dust activity histories, suggests the system has approached something close to a dynamic steady state rather than still accumulating toward an unknown ceiling. For the process marker interpretation, this is not a complication ,a steady-state concentration is more stable and predictable across the planet than one that varies with the precise balance of local production history. It does, however, mean that the depth profile of perchlorate concentration should reflect the changing production-destruction balance with depth: radiolysis from galactic cosmic rays is most intense near the surface and decreases with overburden, while triboelectric production operates at the surface and delivers perchlorate downward by diffusion and transport. The richest and most radiologically undisturbed perchlorate inventory may therefore sit not at the very surface but a few centimetres below it ,a detail that matters for both the resource architecture and the biosignature preservation argument, and one that targeted depth profiling could resolve.

Chapter 11: Implications for Habitability Assessment

Habitability assessment in Mars science has developed largely as a discipline focused on the past. The frameworks, criteria, and vocabulary of the field ,water activity, energy sources, nutrient availability, pH and temperature compatibility with known metabolic pathways ,are tools for evaluating whether ancient environments could have supported life. This focus is not arbitrary. Mars presents an ancient aqueous record in its mineralogy and stratigraphy, and if life ever arose there, it most likely did so during the Noachian and early Hesperian when liquid water was widespread and geochemical energy sources were active. Reading that record is a legitimate and important scientific enterprise.

The complication is that habitability in the past and detectability in the present are not the same thing, and the two have been conflated in ways that create systematic blind spots in how exploration is planned. A site can score high on every ancient habitability metric ,prolonged neutral-pH water, abundant redox chemistry, appropriate temperature ranges, preservation in fine-grained sediment ,and still be a poor target for biosignature detection if the materials have been exposed to sustained oxidative processing since the habitable period ended. Conversely, a site with more modest ancient habitability indicators might preserve what biological chemistry was present far better if it has been protected from atmospheric contact by burial or by geographic location in a low-activity zone. Recognizing this distinction forces a more careful separation of two questions that mission planning has tended to treat as one.

The first question ,where was Mars habitable in the past ,is answered by geological and mineralogical evidence. Clay minerals, carbonates, and certain sulfate assemblages indicate past aqueous conditions. Sedimentary textures record depositional environments. Elemental compositions constrain the chemistry of waters that deposited or altered the rocks. This evidence is real, carefully collected, and well-interpreted by decades of remote sensing and rover investigation. Jezero Crater's delta, the clay-bearing units of Gale Crater's Murray formation, the ancient hydrothermal systems visible in orbital spectroscopy across the Nili Fossae region ,these represent genuine records of past habitability that rightly attract scientific attention.

The second question ,where is evidence of past habitability preserved well enough to be detected ,is answered by a combination of the geological record and the post-depositional history of the materials, including the modern atmospheric chemistry that has been operating on whatever the surface currently exposes. This second question has been given far less systematic attention than the first, and the result is a site selection process that optimizes for ancient geological quality while treating preservation quality as a secondary consideration or simply assuming it is adequate at sites with strong ancient indicators.

Modern chemistry enters this picture through the oxidative processing mechanisms discussed in earlier chapters, but it is not the only post-depositional process that matters for preservation. Diagenesis ,the physical and chemical transformation of sediments as they compact and lithify ,alters original compositions in ways that can destroy, modify, or in some cases concentrate biosignature material. Hydrothermal fluid flow through lithified rock transports elements and can oxidize or reduce mineral phases along flow pathways. Impact events excavate and shock-process material, sometimes creating new mineral phases and sometimes thermally destroying organic compounds in the vicinity of the impact. These processes are ancient, and their effects on preservation have been debated in the context of Earth's oldest rocks for decades. On Mars, they are layered beneath the more recent and ongoing effect of surface oxidative chemistry, which is the most operationally relevant process for materials sampled at the present-day surface because its intensity varies spatially in ways that can be characterized with current orbital tools.

The implication for how negative results should be interpreted deserves particular attention, because the question of whether negative results constrain the biology question is central to how the Mars exploration program evaluates its own progress. When a rover samples ancient sediments from a site with excellent habitability credentials and detects no organic matter, or detects organics at concentrations far below what meteoritic delivery alone should have provided, three explanations are available. Life never emerged at this site. Life emerged but left no detectable traces even when the material was fresh. Or life left traces that have since been destroyed or reduced below detection limits by post-depositional chemistry. Without independent information about preservation quality at that site, distinguishing between these explanations is not possible. A negative biosignature result from a well-preserved site carries substantially more weight as evidence against ancient life than a negative result from a poorly preserved site. Reporting negative results without discussing preservation context gives them false equivalence.

This matters for the overall strategy of the Mars astrobiology program because the number of missions available to sample Mars is finite and the time required to mount each one is long. If some fraction of those missions land at sites where biosignature evidence has been degraded by oxidative processing regardless of whether life existed, the program accumulates negative results that do not constrain the biological question as tightly as they appear to. The preservation gradient framework does not change the conclusion that a negative result at a well preserved site would be highly significant ,it strengthens the case for ensuring that at least some of the high-priority missions sample sites where preservation quality can be independently assessed as favorable, rather than assuming that ancient geological quality is a sufficient proxy for preservation quality.

The modern chemistry perspective also opens a supplementary line of inquiry that has barely been developed in Mars mission planning: the possibility that the modern subsurface harbors habitable environments. Perchlorate brines at accessible depths, discussed in Chapter 7, represent the most discussed candidate for present-day liquid water environments, though direct evidence for their existence is lacking and the recurring slope lineage that once seemed to support them have been largely reinterpreted in terms of dry granular flows. The more conservative version of this question does not require liquid water in the present. It asks whether the chemistry of the modern subsurface ,the redox gradients, the mineral assemblages, the trace amounts of water vapor cycling in and out of the regolith ,creates local microenvironments where chemical energy is available and where the most stress-tolerant life forms known from Earth could theoretically maintain slow metabolic activity. This is a genuinely open question, not a claim that modern life exists on Mars. Terrestrial analogs in hyperarid cold deserts, in subsurface basalt aquifers, and in highly saline and oxidizing brines contain microbial communities surviving under conditions that seemed implausible before they were discovered. Whether anything comparable exists on Mars cannot be answered without measurement strategies designed to detect it.

Searching for modern Mars life requires fundamentally different measurement approaches than searching for ancient biosignatures. Ancient biosignature detection relies on chemical and morphological evidence preserved in rock ,molecular patterns, isotopic signatures, and mineral textures. Detecting active metabolism requires looking for chemical disequilibria that would not persist in a purely abiotic system: unexpected atmospheric trace gases produced faster than known abiotic sources would generate them, redox chemistry in near-surface materials that reflects biological electron transfer rather than simple equilibrium chemistry, or time-variable signals ,seasonal changes in gas concentrations, spatial patterns in oxidation state that track temperature or humidity cycles ,that suggest a responsive, metabolically active system rather than passive geochemistry. The methane detections reported from Curiosity's Tunable Laser Spectrometer ,sporadic spikes in atmospheric methane concentration above a low background ,remain unexplained and contested, but they illustrate the category of measurement that would be relevant to this question: chemistry that is surprising and variable in ways that static geochemical models do not predict. Whether those methane observations reflect biology, subsurface geochemistry, or instrumental artifacts is not yet resolved, and they are mentioned here not as evidence for modern life but as an example of the kind of anomaly that a systems-aware habitability framework would treat as a priority for follow-up investigation rather than as a measurement to be explained away.

For human missions, the habitability question shifts again to encompass a pragmatic dimension alongside the biological one. A crew on Mars needs to understand not just whether Mars could support terrestrial life in principle but what the chemical environment actually does to the materials, equipment, and biological systems that humans bring with them. Perchlorate in regolith dust is a chronic exposure hazard if it enters habitats. Oxidizing chemistry in the soil attacks rubber seals, polymer components, and biological tissue. The same atmospheric electrochemistry that produces perchlorate also charges surfaces in ways that can complicate electronic systems and attract dust to critical equipment. Understanding and mitigating these effects requires treating the modern chemical environment as an engineering design constraint ,one that the habitability assessment framework should address as seriously as it addresses water availability and radiation dose.

The integration that the habitability picture ultimately requires is not simply adding a modern chemistry layer to an existing ancient geology framework. It requires recognizing that the two are coupled: modern atmospheric chemistry acts on ancient geological materials, the spatial pattern of that action varies with conditions that ancient geology in part determines (rock permeability, for example, is a geological property with modern chemical consequences), and the resulting preservation quality is the convolution of both ancient history and modern processing that any landed mission actually samples. A habitability assessment framework that keeps ancient and modern as separate categories will systematically misread sites where the interaction between them is what dominates the actual sample chemistry. Getting the framework right matters because the question it is trying to answer is ,was Mars ever inhabited, and if so, can we tell ,is one of the most consequential questions in planetary science.

Chapter 12: The Systems Perspective

A recurring theme in the preceding chapters has been the inadequacy of studying Mars phenomena in isolation. Atmospheric scientists characterize dust devils as boundary layer vortices without asking what those vortices do to surface chemistry. Geochemists measure perchlorate concentrations without systematically connecting them to the dust activity history at the measurement site. Astrobiologists evaluate ancient habitability without accounting for post-depositional modification by ongoing atmospheric processes. Each discipline produces real and important knowledge. The problem is not the quality of the individual contributions but the gaps that open between them ,gaps where some of the most consequential processes on Mars actually operate.

The connection between triboelectric dust charging and perchlorate production is a clear example of how those gaps close when researchers look across disciplinary boundaries. The physics of grain charging in low-pressure granular flows was studied as an atmospheric and aeolian phenomenon. The detection of perchlorate at the Phoenix landing site was interpreted initially as a geochemical or photochemical result. The laboratory experiments that connected these two observations ,demonstrating that electrical discharge in simulated Martian dust-devil conditions produces perchlorate at rates far exceeding those achievable through UV photochemistry ,emerged from groups willing to ask whether the atmospheric physics and the surface chemistry were related rather than treating them as separate problems. That question turned out to change the interpretation of perchlorate from a static legacy of ancient processes to an active product of ongoing atmospheric work. The insight was not exotic; it required the existing knowledge base of both fields. What it required additionally was someone asking whether the two fields were connected, and pursuing the answer experimentally.

A systems perspective formalizes this kind of cross-domain questioning into a methodological stance. It asks not what a given phenomenon is but how it connects to the rest of the system: what drives it, what it in turn drives, what feedback constrain its intensity, and how it has evolved through time as boundary conditions changed. Applied to Mars, this approach reveals a cascade of connections running from solar energy input, through atmospheric thermodynamics, through dust transport and triboelectric charging, through surface chemical reactions, and through the resulting modification of rock and regolith that the next round of dust transport will interact with. The cascade has run for billions of years. The rocks at the Martian surface today are the cumulative output of all of it, and understanding what they record requires understanding the cascade, not just characterizing the endpoint.

One of the practical gains from systems thinking is that it identifies which measurements carry the most leverage ,where a single well-designed observation resolves the most uncertainty across the greatest number of connected questions. A systematic comparison of perchlorate concentrations at sites spanning the documented range of dust activity is one such measurement. It tests the production mechanism, constrains the rate of spatial variability in oxidant production, informs preservation gradient predictions, and provides ground truth for resource assessment models, all simultaneously. The reason this measurement has not been made is not that it is technically difficult ,perchlorate detection is within the capability of instruments already deployed on Mars ,but that mission site selection has not been designed to span the range of atmospheric activity conditions that would make the comparison meaningful. Systems thinking makes the value of that comparison obvious; reductionist thinking, focused on maximizing the ancient geological quality of each individual mission site, obscures it.

The temporal dimension of the systems perspective is particularly important and often underappreciated. Mars has not been operating in its current state for its entire history. The planet was substantially warmer and wetter in the Noachian, with a denser atmosphere, active volcanism, and widespread liquid water. The transition to current conditions was neither instantaneous nor linear ,it unfolded over hundreds of millions of years, with the atmosphere thinning gradually, volcanic activity declining, and liquid water becoming episodic and eventually rare. The dust-driven electrochemistry described in this book operates under current conditions, meaning a thin atmosphere with low pressure and low humidity that maximizes triboelectric charging efficiency and minimizes charge dissipation. Under the denser, wetter Noachian atmosphere, the same triboelectric mechanism would have operated less efficiently because higher atmospheric pressure and greater humidity would have dissipated charge more rapidly. Perchlorate production rates in the early Martian atmosphere may have been significantly lower than they are today. This does not change the conclusion that modern atmospheric chemistry is actively processing surface materials ,it means that the history of that processing is not uniform through time and that materials from different geological periods may carry different cumulative oxidative signatures depending on when they were exposed relative to the transition to present conditions.

Mars orbital parameters add another layer of temporal variation. Axial tilt, orbital eccentricity, and the precession of the orbit vary over timescales of tens to hundreds of thousands of years in ways that alter the geographic distribution of solar insolation and consequently the pattern of surface heating, dust lifting, and atmospheric transport. During periods of higher obliquity ,when the axial tilt was greater than today's roughly 25 degrees ,polar regions received more summer solar energy and equatorial regions received less, shifting the thermal structure of the atmosphere and likely altering both dust storm patterns and regional dust devil activity distributions. The current spatial pattern of dust activity, mapped by Context Camera surveys over the past two Martian decades, represents a snapshot of one particular orbital configuration. Regions that are currently high-activity zones may have been relatively quiet during low-obliquity periods in the recent geological past, and vice versa. For preservation assessment, this means that the cumulative oxidative dose at any location reflects a time-integrated average over orbital cycles rather than simply the current activity level multiplied by exposure duration. Climate models that track orbital variation over the past tens of millions of years are available and could, in principle, be used to compute more accurate cumulative dose estimates than those based on present-day activity alone ,though the uncertainty in such calculations is substantial.

The systems perspective also clarifies the relationship between Mars as a subject of scientific study and Mars as a destination for human activity. These two framings are sometimes treated as separate concerns ,science happens before humans arrive, and engineering takes over after ,but the systems view shows why that separation is artificial. The same atmospheric processes that matter for biosignature preservation also matter for human engineering: dust accumulation on solar panels, perchlorate exposure to crew health, triboelectric charging of equipment surfaces, and seasonal variation in power generation all connect back to the dust-active, electrochemically productive atmosphere that the science chapters describe. The resource questions ,where to find water, oxygen, and raw materials ,are answered by the same characterization of surface chemistry and atmospheric processing that the astrobiology questions require. A systems framework that describes how Mars actually operates serves both scientific and engineering purposes simultaneously, which is the most efficient use of the limited observational data available from a planet where every measurement is expensive.

What the systems perspective ultimately demands is a different relationship between mission design and scientific framework. Current practice tends toward sequential logic: identify the best ancient geological site available, land there, and analyze what is found. The scientific framework informs the instrumentation and the interpretation but rarely drives the fundamental choice of where to go and what spatial comparisons to build in. A systems-aware approach inverts part of this logic: the framework predicts that certain spatial comparisons ,high-activity versus low-activity sites, deep versus shallow samples, recently exhumed versus long-exposed surfaces ,will resolve the questions that matter most for both science and eventual human operations, and mission design is built around enabling those comparisons rather than around maximizing the geological credentials of a single site. This is not a criticism of how Mars science has been conducted to date. The program has made real and important progress under the existing paradigm. It is an argument that the next generation of questions ,about preservation gradients, about the activity-chemistry link, about the depth structure of oxidative alteration ,requires a different mission design philosophy to answer efficiently. Systems thinking is what makes that philosophy legible.

Chapter 13: Testing the Framework

A framework that cannot be tested is not science ,it is narrative. The dust-driven chemistry framework developed in this book makes specific, falsifiable predictions about spatial patterns in Martian surface composition, the relationship between atmospheric activity and oxidant abundance, the depth structure of chemical alteration, and the resulting variation in organic preservation across sites with different exposure histories. These predictions are not vague directional claims. They are quantitative enough that measurements designed to test them should be able to return a clear answer about whether the framework holds, needs revision, or should be abandoned in favor of a better explanation. Laying out those predictions explicitly, and identifying which near-term missions and measurement strategies are capable of testing them, is what distinguishes a working scientific hypothesis from an interpretive story.

The most direct and consequential test is comparative perchlorate measurement across sites selected specifically to span the documented range of dust devil activity. The prediction is that sites with higher cumulative dust activity should carry higher perchlorate concentrations in surface and near-surface materials, assuming comparable geological settings and exposure ages to control for confounding variables. The prediction does not require order-of-magnitude differences;atmospheric mixing over geological time would tend to reduce extreme gradients ,but it requires that the differences be systematic and detectable above measurement noise if the triboelectric production mechanism is the dominant source. What makes this test demanding is the site selection requirement: it cannot be done by re-analyzing existing measurements from Phoenix and Curiosity, both of which sit at moderate-activity sites that provide no contrast. It requires at minimum one landing in a high-activity zone and one in a low-activity zone, with identical analytical protocols applied to equivalent sample types at both. No currently operating mission has been designed with this comparison as an explicit objective, and no planned mission as of writing has made it a primary driver of site selection.

Mars Sample Return, currently the highest-priority large mission in the NASA planetary science portfolio, offers a partial path toward this test but does not fulfill it on its own. Samples cached by Perseverance at Jezero Crater will provide extraordinarily detailed chemistry from a single location. When those samples eventually reach terrestrial laboratories, they can be analyzed with instruments and techniques impossible to fly ,perchlorate isotopic composition, complex organic molecular structure, micrometer-scale mineral texture, and trace element distributions that no rover instrument resolves. This is enormously valuable. But Jezero is one site, with one dust activity history, and the comparison the framework requires demands at least a second site with a substantially different activity history. The scientific return from Mars Sample Return would be substantially amplified if a second sample return mission targeted a site chosen partly on the basis of contrasting activity level ,not instead of geological interest, but alongside it. The argument for doing so is not subtle: the samples from Jezero will be interpreted against whatever assumptions are made about preservation quality at that site, and those assumptions currently rest on very limited data. A second site with a characterized preservation history would transform those assumptions into testable constraints.

Depth profiling is a second critical test that no current Mars surface mission is equipped to perform adequately. The oxidation penetration model developed in Chapter 5 predicts that surface materials at an exposed site should be more oxidized than subsurface materials at comparable depth, with a transition zone whose depth scale reflects the local combination of dust activity intensity, rock permeability, and surface exposure age. Testing this prediction requires sampling the same geological unit at multiple depths within a single borehole, with a depth range reaching at least tens of centimeters and ideally one to several meters. The best-case outcome would be a clear oxidation gradient matching the predicted depth scale ,more oxidized mineralogy and lower organic abundance near the surface, grading toward less altered material at depth. An absence of depth-dependent variation would indicate either that penetration depths are shallower than modeled, making surface samples adequate after all, or that the timescales of equilibration are short enough that even subsurface material has reached near-surface oxidation levels at this particular site. Either result is informative. Neither can be obtained without drilling capability substantially beyond what Perseverance or Curiosity possesses. The InSight lander's failed subsurface heat flow probe is a cautionary reminder that deep drilling on Mars involves mechanical challenges that laboratory testing does not always anticipate, and any future deep-drilling mission design will need to account for regolith properties that can vary significantly from orbital predictions.

Organic abundance variation across sites with different predicted preservation quality constitutes perhaps the cleanest test the framework offers. If modern oxidative processing is a significant control on how much organic material survives at the Martian surface, then sites with low cumulative dust activity and short surface exposure should retain more organic carbon ,from meteoritic delivery if nothing else ,than sites with high activity and long exposure. Meteoritic organic delivery to Mars is estimated at roughly ten to a hundred nanograms per gram of surface material accumulated over geological time. Current Sample Analysis at Mars detections of organics in Gale Crater mudstones run at the parts-per-billion level by mass, substantially below what meteoritic delivery alone should have provided even without any ancient biogenic contribution. This deficit is consistent with ongoing oxidative destruction of organics, but proving that interpretation requires showing that sites predicted to have better preservation actually retain more organic material. A multi-site comparison, controlled for geological context, that found organic abundance correlating positively with predicted preservation quality would be one of the strongest results Mars exploration could produce ,not because it would prove ancient life but because it would validate the preservation framework that determines how life detection results should be interpreted at any given site.

Seasonal variation in surface chemistry provides a shorter-timescale test that is accessible to long-duration landed missions without requiring new landing sites. If perchlorate production is ongoing through triboelectric activity in dust devils, then surface material at a moderately active site should show small but detectable changes in oxidant levels between the active dust devil season and the quiet season. The Curiosity rover has operated long enough to sample multiple Martian years and has documented clear seasonal patterns in atmospheric chemistry including methane variability and water vapor cycling. Applying the same multi-year time series approach to surface perchlorate measurements ,sampling the same material type at the same location across seasons ,would test whether chemistry responds to atmospheric activity on annual timescales. The signal is expected to be small, potentially below measurement uncertainty for individual measurements, but statistical analysis across multiple seasonal cycles could reveal it if it exists. This test requires commitment of rover resources to repeated sampling at previously analyzed locations, which competes with the scientific imperative to investigate new terrain. Mission operations teams would need to treat it as a scientific priority rather than an operational constraint.

The framework's predictions about subsurface brine stability ,discussed in Chapter 7 ,represent a third category of testable claim, though one requiring measurement capabilities not yet deployed at Mars. Ground-penetrating radar at sufficient resolution to detect centimeter-scale liquid layers, direct thermal and humidity profiling through the top meter of regolith, or direct chemical sampling at depths of tens of centimeters with phase identification capability would all constrain whether perchlorate brines form and persist under the temperature and humidity conditions that models predict should support them. The lack of this measurement capability is not a scientific failure ,it reflects genuine engineering constraints on what can be achieved within current mass and cost budgets for Mars surface missions. But it means the brine component of the framework remains a prediction without a clear test pathway in the near-term mission queue, and the planning community should treat this as a gap worth addressing rather than as a settled question in either direction.

What gives the framework credibility despite these gaps is not that it is consistent with existing observations ,many hypotheses can be constructed to accommodate the same limited dataset. It is that the predictions are specific enough to be distinguished from alternatives by measurements that are achievable with realistic near-term mission concepts. A hypothesis that is consistent with all possible outcomes of all conceivable tests is not scientifically useful. A hypothesis that would be clearly contradicted if perchlorate concentrations showed no spatial structure across sites with very different activity histories, or if organic abundance were uniformly low regardless of predicted preservation quality, is one that can actually advance understanding by either being confirmed or being ruled out. The dust-driven chemistry framework earns its place in the scientific conversation by meeting that standard.

Chapter 14: Resource Implications

The resource picture for Mars has traditionally been assembled from geological inventory: where is water ice, how much CO_2 is in the atmosphere, where do minerals with useful elemental compositions outcrop, what is the estimated depth and extent of specific deposits. This inventory approach is appropriate for resources that exist because of static geological history and sit in specific locations waiting to be found and extracted. It is less well-suited to resources that are continuously produced and redistributed by ongoing atmospheric processes, because those resources do not have a static distribution ,they have a production rate, a transport pattern, and accumulation zones determined by current and recent dynamics rather than ancient geology alone.

Perchlorate is the clearest example of a resource governed by ongoing production rather than static inventory. Its global distribution at roughly half a percent by weight across geologically diverse sites reflects continuous atmospheric manufacturing and redistribution, not the remnant of a specific ancient depositional event. This distinction matters for extraction strategy in a practical way. An inventory-based approach would prospector for enriched perchlorate deposits the way one would search for ore bodies ,identifying locations where ancient processes happened to concentrate the material above background levels. A production-based approach would instead characterize where current and recent atmospheric processes deliver the highest supply rate, positioning extraction infrastructure at those locations to intercept material as it accumulates rather than mining a deposit that may or may not extend to depth. The two approaches are not mutually exclusive, but the second one is only visible if you start from the recognition that Mars is actively producing the material rather than simply holding a static inventory of what ancient processes left behind.

The oxygen implications of perchlorate abundance deserve careful quantification rather than qualitative optimism. Thermal decomposition of calcium perchlorate yields approximately 0.26 kilograms of oxygen per kilogram of perchlorate processed. At a surface concentration of 0.5 percent by weight in bulk regolith, extracting and processing one metric ton of regolith yields approximately 1.3 kilograms of oxygen ,enough to supply one person's breathing requirement for a little over a day. These numbers are not discouraging on their own, but they establish the scale of processing required if perchlorate is the primary oxygen source. A crew of four consuming roughly four kilograms of oxygen per day for breathing alone ,not counting propellant oxidizer needs, which would increase requirements by orders of magnitude for a return mission ,would require processing approximately three tons of regolith daily just to meet the breathing requirement through perchlorate alone. This is within reach of a well-designed excavation and thermal processing system, but it requires acknowledging the throughput requirement explicitly rather than treating perchlorate as an effortlessly abundant oxygen source. MOXIE, the Mars Oxygen In-Situ Resource Utilization Experiment carried on Perseverance, demonstrated atmospheric CO_2 electrolysis as a complementary oxygen production pathway, converting CO_2 to oxygen and carbon monoxide. For a settlement, the optimal approach likely combines both pathways depending on available power and the specific operational phase ,atmospheric electrolysis for baseload production where power allows, perchlorate decomposition as a batch process when high-purity oxygen in concentrated form is needed and bulk atmospheric processing is less practical.

Water extraction from Martian sources involves a hierarchy of options with different mass efficiencies, energy costs, and geographic constraints. Near-surface water ice at high latitudes ,confirmed by Phoenix, by SHARAD radar sounding, and by orbital neutron spectroscopy showing elevated hydrogen near the poles and at mid-latitudes ,is the most mass-efficient water source where it exists at accessible depths. Processing ice into potable water requires heating to sublimate or melt it and condensing the vapor, an energy cost that scales with the thermal properties of the ice-regolith mixture and the efficiency of the condensation system. At sites where near-surface ice is unavailable ,lower latitudes where ice is unstable at shallow depths under current thermal conditions ,alternatives include atmospheric water vapor capture, extraction from hydrated minerals, and extraction from perchlorate-bearing hygroscopic regolith. Atmospheric water capture is technically straightforward using sorbent materials that adsorb water vapor during cold nights and release it when heated, but the low atmospheric water vapor abundance ,typically 10 to 100 precipitable micrometers depending on season and latitude ,means that large volumes of atmosphere must be processed for modest water yields. Hygroscopic regolith extraction, where perchlorate salts have concentrated atmospheric moisture in the shallow subsurface, offers water yields per unit mass processed that are intermediate between ice extraction and pure atmospheric capture, but the actual water content per kilogram of regolith depends heavily on local perchlorate concentration and the humidity history of the site ,quantities that require in-situ measurement to constrain.

The spatial structure of resource availability creates an explicit tension between the site selection criteria that optimize for science and those that optimize for sustained human operations. The ancient lake beds, hydrothermal systems, and sedimentary sequences that represent the highest-priority targets for biosignature science tend to cluster at low to mid latitudes in geologically diverse terrain. High-latitude sites where near-surface ice is abundant and accessible are less geologically interesting for ancient habitability questions but offer the most favorable water resource profile. This tension is not insurmountable ,low-latitude sites with subsurface ice at modest depths may exist, and near-equatorial regions with enriched perchlorate or accessible hydrated minerals may provide water more easily than average regolith at those latitudes ,but acknowledging the tension honestly is preferable to claiming that optimal science sites and optimal resource sites simply coincide. For the first human missions, which will be primarily focused on survival and technology demonstration rather than settlement, the water resource situation at the landing site may dominate site selection. For subsequent missions as the program matures, the balance can shift toward geological and scientific interest. Separating these mission phases conceptually, and designing them with their respective priorities made explicit, produces a more coherent long-term program than trying to satisfy all objectives simultaneously from the outset.

The economic logic of in-situ resource utilization on Mars is governed by the relationship between launch cost and the mass of consumables that resource production can replace. At current and near-future launch costs to Mars ,even with the dramatic reductions that fully reusable vehicles are beginning to enable ,water, oxygen, and propellant are the largest mass drivers for long-duration surface operations and for return missions. Closing these loops locally, even partially, produces substantial leverage on total mission mass. A human Mars mission that produces its return propellant from local CO_2 and water rather than launching it from Earth reduces the initial mass in Earth orbit by a factor that various analyses have placed between five and eight, depending on specific mission architecture. This is the fundamental argument for in-situ resource utilization as a mission enabler rather than a convenience: it is not about saving money at the margin but about making missions architecturally feasible that would otherwise require infrastructure launch capacity that does not currently exist. The Mars Oxygen In-Situ Resource Utilization Experiment's demonstration on Perseverance ,producing oxygen from Martian CO_2 at rates that scaled to mission-relevant throughput would meet crew breathing requirements ,showed that the core chemistry works under actual Mars surface conditions rather than just in laboratory simulations. That demonstration matters because the history of Mars resource concepts includes many laboratory-validated ideas that behaved differently in the actual Martian environment. MOXIE earned its place in mission history by surviving the transition from Earth laboratory to Mars surface.

The picture that emerges is one where resource strategy on Mars rewards a systems-aware approach ,mapping not just where specific materials exist in static deposits but where ongoing processes deliver useful materials at rates high enough to intercept, and designing extraction infrastructure to work with those delivery rates rather than against them. Dust is the most pervasive example: it accumulates on every surface, carries perchlorate and other chemically processed material from across the planet, and represents a resource stream that the atmosphere provides continuously without any excavation. Solar panels already interact with this stream as a contamination problem; thinking of it as a resource stream that could be intercepted and processed is a shift in perspective that costs nothing in mass but potentially yields meaningful returns in oxygen and water feedstock if the collection and processing technology matures sufficiently. Whether that maturation happens on the timescale relevant to early human missions is an engineering development question, not a physics question ,the physics works, and the question is whether the engineering investment to turn it into a reliable operational system will be made before it is needed.

Chapter 15: Mission Architecture Considerations

Mission architecture for Mars is constrained by physics in ways that do not forgive poor planning. The mass of every kilogram delivered to the Martian surface is paid for in propellant, structural mass, and thermal protection upstream. The light-speed communication delay ,ranging from three to twenty-two minutes one-way depending on orbital geometry ,constrains autonomous operation and limits how quickly ground controllers can respond to surface discoveries or contingencies. The eighteen- to twenty-six-month cycle between Earth-Mars launch windows limits resupply and crew rotation frequency in ways that have no analogy in terrestrial logistics. Mission architects working within these constraints make tradeoffs that look very different from the tradeoffs facing engineers designing Earth-based systems, and the framework developed in this book has specific implications for what those tradeoffs should look like.

The most immediate implication concerns drilling capability. Current Mars rovers ,both Curiosity and Perseverance ,are equipped with percussion drill systems that collect cores approximately five to six centimeters in length from a few centimeters below the surface. This capability was designed for geochemical characterization and sample caching, not for depth profiling through an oxidation zone that may extend tens of centimeters to a meter or more. The consequence is that every sample collected by current rovers comes from the zone most affected by modern atmospheric oxidation, regardless of where the rover is located. If the preservation gradient framework is correct, current rovers are systematically sampling the most altered fraction of the rock record at every site they visit. Getting below the altered zone requires either a fundamentally different drilling system or careful selection of natural exposures ,fresh impact craters, recent cliff collapses, active erosion faces ,where the geometry of exposure brings deeper material to the surface without requiring the rover to drill to it. Both approaches have merit, but neither is currently a primary design requirement for Mars surface missions.

A mission designed specifically to test the depth structure of oxidative alteration would need to demonstrate at minimum one to two meters of core recovery in coherent rock at multiple locations within a single site. This is a substantial engineering challenge under Mars mass and power budgets. Terrestrial diamond-wire and rotary-percussion systems routinely drill to these depths, but their mass and power requirements are incompatible with current Mars lander or rover architectures. The engineering development path toward useful Mars deep drilling exists ,lightweight pneumatic systems, wire-line drilling that minimizes drill string mass, ice-augering adaptations for permafrost-bearing regolith ,but it requires sustained development investment rather than the incremental improvements to existing technology that have characterized the Curiosity and Perseverance drill systems. Making deep drilling a mission requirement rather than a desired enhancement would force that investment on a timeline relevant to future mission planning.

Long mission duration amplifies science return in ways that are underappreciated in mission cost-benefit analyses focused on minimum success criteria. A rover that operates for one Martian year captures a single seasonal cycle, which may or may not be representative of average conditions. Three or four years of operation document interannual variability and allow time-series measurements of chemistry at repeated sampling locations that would reveal seasonal or storm-driven changes in surface oxidant levels. Both Curiosity and Opportunity demonstrated that rovers designed for nominal mission durations of one to two years can operate productively for a decade or more when their systems are robust and their power sources permit continued operation. The scientific value added by those extended missions ,Curiosity's detection of organics in Murray formation mudstone, Opportunity's documentation of hematite concretion formation environments, both rovers' meteorological records spanning multiple Martian years ,substantially exceeded what the nominal mission plans would have produced. Designing for longevity, rather than treating extended operation as an unanticipated bonus, is a mission philosophy that rewards the investment in robust engineering.

Multi-site coordination between missions from different agencies represents perhaps the most efficient path toward the spatial comparison datasets that the framework's predictions require. A single mission can characterize one site in detail; understanding the spatial structure of Mars chemistry requires multiple sites with contrasting characteristics, measured with consistent techniques. The current international Mars program ,with NASA, ESA, China's CNSA, and India's ISRO all operating or planning Mars assets ,in principle spans enough geographic and temporal coverage to enable this kind of comparison. In practice, instrument packages differ between missions, calibration standards are not always shared, and site selection decisions are made independently within each agency's own scientific priorities and political constraints. The result is a collection of excellent individual measurements that are difficult to compare quantitatively because the measurement protocols were not designed in coordination. This is not a new problem in planetary science ,the history of lunar geochemistry shows similar fragmentation followed by decades of effort to reconcile measurements made with different instruments under different assumptions ,but it is worth naming clearly because the alternative, coordinating measurement protocols and site selection criteria across international missions before launch, is technically achievable if the political will to do it exists. Science diplomacy is not glamorous, but it can multiply the return on existing mission budgets without requiring additional launch vehicles.

The case for a dedicated atmospheric monitoring orbiter ,one whose primary mission is continuous, systematic dust devil and dust storm tracking across multiple Martian years, with high enough temporal and spatial resolution to characterize individual vortex events and accumulate statistically robust activity maps ,has grown stronger as the Context Camera analysis described in earlier chapters has demonstrated the scientific value of even retrospective activity surveys from an instrument not designed for that purpose. An orbiter optimized for this purpose, carrying a wide-field camera with fast cadence, a thermal imager to map surface heating patterns that predict convective activity, and a radar sounder for near-surface structure characterization, would provide the baseline activity dataset that all surface missions need to interpret their measurements in terms of local processing history. Such a mission would not directly detect biosignatures or collect returnable samples, and it would therefore score relatively poorly on the metrics that dominate prioritization in decadal surveys focused on the most exciting immediate science. But it would substantially amplify the scientific return of every subsequent surface mission that uses its activity maps for site selection and measurement interpretation. The case for infrastructure missions that enable better science from operations missions is structurally sound even when it is politically difficult to advance.

For the eventual transition to human operations, the mission architecture implications of the active chemistry framework converge on a set of practical requirements that flow from the resource and preservation arguments of preceding chapters. Water production infrastructure should be designed to operate across a range of source materials ,atmospheric capture, ice excavation, and hygroscopic regolith processing ,with the flexibility to adjust the mix as in-situ resource characterization reveals what is actually available at the specific site. Oxygen production should plan for both CO_2 electrolysis and perchlorate decomposition as complementary pathways rather than committing to one before understanding the local resource profile. Dust management ,protecting equipment, power systems, and crew health from the dust that is both a resource stream and a contamination problem ,needs to be addressed as a fundamental design constraint rather than a secondary consideration, because the operational history of Mars solar power systems shows that dust accumulation is among the most significant performance degraders in extended surface operations. The Spirit rover's ultimate loss of power was directly caused by dust accumulation on its solar panels; designing human missions that are not similarly vulnerable requires either nuclear power, active dust-removal systems, or both.

Power system architecture underlies everything else, because the resource production processes and the life support systems both require reliable, continuous power in quantities that solar alone cannot guarantee through Mars's global dust storm seasons, when solar flux at the surface can drop to a few percent of clear-sky values for weeks at a time. Small fission surface power systems ,of the type that NASA and the Department of Energy began ground testing in the mid-2020s under the Kilopower and successor programs ,provide the baseload power that dust-resilient human operations require. Their mass, roughly one to two metric tons for a ten-kilowatt system, is significant but not prohibitive in a human mission architecture that already requires tens of tons of landed mass for habitats, life support, and mobility systems. Treating nuclear surface power as a required element of human Mars architecture rather than an optional upgrade is the systems-aware conclusion of the resource and operational analysis ,the alternative is designing a mission whose operational capability collapses during the first global dust storm, which is a known recurring event on a planet where dust storm seasons arrive every Martian year and planet-encircling storms occur at intervals of roughly every three to four Earth years.

The thread connecting all of these architectural considerations is the same thread running through the scientific chapters: Mars is an active system, and mission design that ignores that activity ,whether by sampling only the oxidized surface layer, or by powering operations solely from a sun that dust storms periodically obscure, or by planning resource extraction from static geological deposits without accounting for the ongoing atmospheric production and redistribution of key materials ,will perform worse than designs that account for it. The planet operates by rules that are well enough understood to plan around. Using that understanding is not optional for missions that aim to succeed across the full range of Martian conditions rather than only under the favorable subset.

Chapter 16: Looking Forward

The first phase of Mars exploration answered the broadest questions. The planet has a thin atmosphere composed mostly of carbon dioxide. Its surface carries the mineralogical record of past liquid water in the form of clay minerals, sulfates, and carbonates. Ancient river valleys and lake basins are visible from orbit and confirmed by surface missions. The global magnetic field is absent, the crust records ancient regional magnetism in patches, and the current surface environment is cold, dry, and exposed to ultraviolet radiation. These are settled facts, well-established by five decades of missions.

What is less settled is how to read the surface chemistry of Mars correctly ,not as a static record written once and preserved unchanged, but as the product of two overlapping processes: the ancient geological and aqueous history that produced the mineralogy in the first place, and the modern atmospheric chemistry that has been modifying whatever sits at the surface ever since it was exposed. The case developed throughout this book is that these two processes cannot be cleanly separated at most locations accessible to current missions, and that treating them as separable ,by focusing exclusively on ancient geological credentials when selecting landing sites ,systematically introduces bias into how results are interpreted.

The next generation of Mars missions could, if designed with these interactions in mind, begin closing the key gaps. The most pressing gap is the absence of any coordinated multi-site comparison of surface chemistry across locations with different documented dust activity histories. Every perchlorate measurement made so far on Mars comes from a single analytical campaign at a single site, and both sites where the measurements exist ,the Phoenix landing site in the northern lowlands and Gale Crater where Curiosity has operated ,fall in the moderate-activity portion of the dust activity spectrum. The high-activity zones clearly visible in Context Camera orbital surveys and the quietest ancient highland terrains have never been sampled for surface chemistry with the intent of making a controlled comparison. That comparison is what the framework developed here predicts will show systematic differences in oxidant abundance and organic preservation, and it cannot be made by re-analyzing existing data. It requires mission planning that treats activity level as a deliberate site selection variable rather than an incidental characteristic.

Sample return is the most powerful near-term tool for closing analytical gaps that in-situ instruments cannot bridge, because the measurements most diagnostic for distinguishing biological from abiotic chemistry ,perchlorate isotopic composition, complex organic molecular structure, micrometer-scale mineral texture and redox patterning ,require laboratory instruments too massive and power-hungry to fly. The Mars Sample Return program, centred on Perseverance's cached samples from Jezero Crater, will eventually provide this level of analysis for one location. Jezero is a well-chosen site on ancient geological grounds ,a river delta depositing sediment into a lake basin roughly 3.5 billion years ago, with mineralogy indicating sustained water-rock interaction. What the samples will not tell us, taken alone, is how much of the original chemistry at that site survives intact versus how much has been modified by post-depositional processes including modern atmospheric oxidation. That contextual question requires a second site with a different processing history to answer, because without contrast there is no calibration. A second sample return mission choosing its target partly on the basis of contrasting dust activity and surface age would multiply the interpretive value of the Jezero samples substantially, because each site illuminates the other by providing the comparison the framework requires.

Drilling capability is a hardware gap with direct consequences for the science that current and near-future missions can produce. Curiosity and Perseverance drill to roughly five to six centimetres depth. If the oxidation penetration estimates developed in Chapter 5 are even approximately correct ,active oxidant zones extending from centimetres to perhaps a metre or more depending on rock permeability and exposure age ,then current rovers are sampling almost exclusively within the modified zone at any site with meaningful surface exposure age. Accessing demonstrably unaltered ancient material without drilling requires identifying natural exposures where geometry brings deeper material to the surface: fresh impact craters that have excavated into ancient stratigraphy, active erosion scarps, recently exposed fault faces. These exist on Mars and are worth targeting deliberately. But relying entirely on geological luck to expose the right material at accessible depth is not a systematic strategy. Developing and flying a drill system capable of recovering coherent core from one to two metres depth is an engineering investment that would change what Mars surface missions can actually measure, and it belongs in mission requirements rather than in the category of desirable enhancements.

The international dimension of Mars exploration is a practical reality that affects how quickly multi-site datasets can be assembled. No single agency's budget or launch manifest can sustain the landing frequency that systematic multi-site comparison requires on a reasonable timescale. Coordination ,on measurement protocols, data sharing, and at minimum the publication of raw calibration data in formats that allow cross-mission comparisons ,multiplies the scientific return on existing investments without requiring additional launches. The obstacles to this coordination are political rather than technical, which does not make them easier to solve, but it does mean that the scientific community can articulate clearly what would be gained and leave the negotiating to others. What would be gained is the ability to treat Mars landed missions collectively as a spatially distributed observing network rather than as a collection of independent single-point characterizations. The difference between those two things, for testing spatially structured hypotheses like the preservation gradient framework, is the difference between data that can answer the question and data that can only suggest it.

Chapter 17: A Planet in Transition

The title of this book has two meanings that are worth holding together. Mars is a planet geologically in transition ,not in the dramatic sense of active volcanism or running rivers, but in the quieter sense that its surface continues to change through erosion, impact gardening, aeolian redistribution of material, and the slow chemical modification of exposed rocks by an atmosphere that, while thin, has not stopped working. Ancient lake beds erode grain by grain. Dust redistributes across hemispheres during global storms. Impact craters occasionally excavate fresh material from depth. The planet has not frozen into a final state. It is still moving, slowly, through a long decline from its earlier dynamic period.

Mars science is also in transition, and in some ways the scientific transition is more consequential for exploration than the geological one. The view of Mars as a chemically inert museum ,ancient chemistry preserved intact, waiting for instruments sensitive enough to read it ,has been productive as a working assumption for early missions, because it justified going to the places with the strongest ancient geological signals and looking carefully at what those places contain. That approach produced real results: the detection of organics in Gale Crater mudstone, the characterisation of delta stratigraphy at Jezero, the documentation of widespread clay and sulfate mineralogy indicating past aqueous environments across broad regions of the planet. These are genuine scientific achievements, not artefacts of a flawed paradigm.

The limitation is not that the museum view was wrong to focus on ancient geology. The limitation is that it did not ask what has happened to the ancient geology since it formed. The answer, developed across the preceding chapters, is that the answer varies by location and by depth in ways that the framework here makes physically interpretable and, in principle, measurable. Some ancient materials sit at the surface in regions where dust devil activity has been high for geological timescales and where the cumulative oxidative dose delivered to exposed rock has been large. Other ancient materials sit in quieter regions, or are buried under younger cover that has only recently been breached by erosion, or occupy depth intervals below where atmospheric oxidants penetrate efficiently. These materials have experienced very different post-depositional histories, and that difference is written into the chemistry of what they contain today, even though the ancient environments that deposited them may have been comparable.

Recognising this is not the end point of a scientific argument ,it is the starting point of a measurement programme. The framework makes predictions that are specific enough to test: perchlorate concentrations should vary systematically with dust activity level across sites with controlled geological context; organic abundance should correlate with predicted preservation quality; depth profiles should show oxidation gradients whose scale reflects local activity and permeability. None of these predictions has been tested in a way that was designed to test them. They have been consistent with the observations made so far, which is a necessary but not sufficient condition for confidence. Confirmation requires the targeted measurements that designed tests produce, and those tests have not yet been designed into the mission queue.

This is not a criticism of the Mars science community, which has produced extraordinary work under genuine constraints of cost, mass, and political will. It is an observation about where the frontier of understanding currently sits and what would move it. The reconnaissance era of Mars exploration ,characterising basic properties at individual locations ,has done its job. The systematic era, testing process hypotheses across spatial gradients and comparing results between sites chosen to span the relevant variable space, is the logical next step. It requires a different design philosophy: missions built around comparisons rather than around single-site characterisation, with site selection driven partly by the need to span the range of modern processing conditions rather than purely by the quality of ancient geological indicators.

The dust-driven chemistry framework is one piece of a larger systems picture that Mars science has been assembling for decades. Atmospheric dynamics connect to surface chemistry through dust transport. Surface chemistry connects to subsurface conditions through oxidant diffusion and periodic brine cycling. Subsurface conditions connect back to the surface through what erodes and what remains buried. Each connection opens new questions and suggests new measurements, and the connections are real ,they are grounded in physics and chemistry that are not in dispute, even where the quantitative magnitudes of specific effects remain poorly constrained. Working through those connections, rather than treating each subdiscipline as self-contained, is what turns Mars from an object of description into a system that can be understood and eventually worked within.

The ancient questions remain open. Whether Mars was ever inhabited is not known, and current evidence does not settle it. The absence of confirmed biosignatures after multiple missions to ancient aqueous terrains is consistent with life never having arisen, but it is also consistent with life having arisen and left traces that have since been degraded, or with traces existing at locations and depths not yet sampled. The framework in this book makes the second interpretation more plausible than it would be under a uniform-preservation assumption, because it identifies physical reasons why the surface chemistry at accessible locations would be expected to have been modified since deposition. It does not make the second interpretation certain ,it makes it a scientifically serious possibility that warrants targeted investigation rather than dismissal.

Part Two: The Active Resource

From Atmospheric Chemistry to In-Situ Utilisation

Chapter 18: Dust as a Resource Stream

Every Mars mission that has used solar power has contended with dust. The Spirit rover's operational life ended in 2010 when dust accumulation on its panels reduced power below the threshold needed to survive a Martian winter. Opportunity survived longer because wind-cleaning events periodically cleared its panels, extending its operational life well beyond design expectations. The pattern those two missions established ,dust is the primary long-term threat to solar-powered surface operations ,has shaped every subsequent mission design discussion about power systems on Mars.

Dust is a genuine operational problem, and nothing in what follows is intended to minimise that. But the engineering response to dust has been almost exclusively defensive: design for dust tolerance, build in cleaning mechanisms, monitor panel output, plan for power degradation. The question worth asking alongside the defensive one is whether the material that accumulates on surfaces has any value ,whether the dust that arrives continuously, pre-sorted by size through atmospheric processes and pre-charged through triboelectric interactions in the same dust devils that drive the electrochemistry this book has been describing, constitutes a resource stream that could be intercepted rather than simply excluded.

The composition of Martian airborne dust is reasonably well-characterised from orbital spectroscopy and from in-situ measurements by multiple landed missions. Fine-grained dust is dominated by iron oxides, mafic silicates, and salts including sulphates and perchlorates. The global uniformity of dust composition ,its spectral properties are nearly identical regardless of where on the planet it is measured ,reflects the efficiency of atmospheric mixing during dust storms, which transport fine particles globally and homogenise the composition of the smallest size fraction over geological time. The perchlorate content of fine dust is consistent with the half-percent-by-weight concentrations measured at Phoenix and Curiosity, meaning that every kilogram of fine dust that settles on a surface carries roughly five grams of perchlorate alongside iron oxide and silicate minerals.

The surface area of any Mars installation presents a passive collection target to this dust flux. A habitat with a footprint of a hundred square metres, combined with solar panels, antennae, and surrounding cleared ground, presents several hundred square metres of surface to the atmosphere. Measured dust deposition rates on Mars surfaces vary by season and location but fall roughly in the range of one to ten milligrams per square centimetre per year in moderate-activity regions, based on estimates derived from solar panel power degradation rates and from direct mass measurements where available. At the lower end of that range, several hundred square metres of collection surface would accumulate on the order of a few kilograms of dust per year. At the higher end, tens of kilograms per year. These are rough estimates, and actual rates at a specific site would need to be measured in situ;the numbers are offered here as order-of-magnitude context rather than design specifications.

The value of that material depends on what fraction is perchlorate and what the perchlorate is worth in terms of oxygen production. At five grams of perchlorate per kilogram of dust, and with perchlorate yielding approximately 0.26 kilograms of oxygen per kilogram processed, a kilogram of dust contains roughly 1.3 grams of extractable oxygen. Ten kilograms of passively accumulated dust per year would yield roughly 13 grams of oxygen ,a trivially small quantity relative to the roughly 1.5 kilograms per day that a single person requires for breathing. Passive collection from existing surfaces alone, at these deposition rates, does not come close to constituting a primary oxygen source.

This arithmetic is worth doing explicitly because the concept of dust-as-resource is sometimes presented with more optimism than the numbers support. Passive collection from incidentally available surfaces is not a meaningful contributor to crew oxygen supply. The concept becomes potentially relevant only if the collection area is deployed deliberately and at scale ,hundreds or thousands of square metres of surface specifically designed and positioned to maximise dust capture ,and only if the collected material can be concentrated and processed efficiently. Deploying large collection arrays adds landed mass, and the mass-efficiency of that approach compared to simply landing more oxygen or more MOXIE-equivalent electrolysis capacity is an engineering trade that has not been resolved by any published analysis. The honest assessment is that dust collection as a primary resource strategy requires demonstration and quantification before it can be planned around.

Where the concept has more defensible near-term value is as an adjunct to infrastructure that will be deployed for other purposes. Solar panels will be cleaned periodically regardless of any resource intention. If the cleaning mechanism captures the removed dust rather than dispersing it, the collected material costs essentially nothing in additional mass beyond the collection hopper. The dust then becomes available for analysis and, if processing infrastructure exists, for extraction of perchlorate and adsorbed water. This incremental approach ,harvesting what is already being removed as a maintenance task ,is a lower-risk starting point than purpose-built large-area collection arrays, and it would generate real data on collection rates, composition, and processing efficiency under actual Mars surface conditions.

Dust composition varies with season in ways that have direct implications for any collection strategy. During the global dust storm season, when planet-encircling storms loft material from across the planet into suspension and redistribute it globally, the composition of settled dust reflects the global average of the Martian surface. During quiescent periods, local wind regimes transport material from nearby source regions, and collected dust may show compositional signatures of local geology. For perchlorate specifically, this seasonal variation matters if high-activity dust devil seasons produce freshly processed perchlorate-bearing particles that settle out in greater quantity during those periods. Measuring whether settled dust perchlorate content varies seasonally, using instruments already present on a well-instrumented lander or rover, would test this prediction directly and cost no additional mission mass.

Chapter 19: Perchlorate Processing Units

The case for using Martian perchlorate as an oxygen source rests on chemistry that is well-understood and that has been demonstrated in the laboratory under Mars-relevant conditions. The basic reaction is not in dispute: heating calcium or magnesium perchlorate above roughly 400 degrees Celsius causes thermal decomposition, releasing molecular oxygen and leaving behind the corresponding metal chloride. For calcium perchlorate specifically, stoichiometry yields approximately 0.26 kilograms of oxygen per kilogram of perchlorate, with calcium chloride as the solid residue. This is straightforward thermochemistry, and the reaction proceeds reliably once the temperature threshold is met and maintained.

The relevant questions for Mars application are not whether the reaction works but how efficiently a processing system can collect feedstock, maintain decomposition temperature under Mars thermal conditions, capture and store the released oxygen, and handle the residual chloride ,and what the mass and power cost of doing all of that is relative to the oxygen quantity produced. These are engineering questions whose answers depend on specific system designs, and the honest answer is that no full system has been built, tested, and characterised under Mars surface conditions. What exists is laboratory validation of the core chemistry, conceptual system designs, and thermal and energy models whose predictions have not been confirmed by operational hardware.

Thermal management is the first practical challenge. Mars surface temperatures range from roughly 150 Kelvin at polar night to around 300 Kelvin at equatorial midday. A reactor operating at 400 degrees Celsius ,673 Kelvin ,sits several hundred degrees above even the warmest ambient conditions, meaning substantial heat input is required to initiate and sustain decomposition and substantial insulation is required to prevent that heat from bleeding away into the environment. On Earth, insulating a reactor to maintain 400 degrees Celsius against an ambient of 20 degrees Celsius is standard industrial practice. On Mars, the same reactor faces 150-degree ambient at worst case, which is actually a smaller temperature differential ,but the Martian atmosphere's low pressure reduces convective heat transfer, which can be an advantage for retaining heat within a well-designed system. The engineering is tractable. It requires careful thermal design and materials selection, but it does not require any physics beyond what terrestrial high-temperature systems routinely manage.

Electrochemical decomposition of perchlorate is an alternative that operates at lower temperature but introduces a different set of engineering requirements. Electrochemical cells for perchlorate reduction have been demonstrated in the laboratory at reasonable energy efficiencies, but they require electrolyte solutions that must remain liquid and ionically conductive at Mars-relevant operating temperatures, electrode materials stable under sustained oxidising conditions, and power delivery systems compatible with the cell's current-voltage characteristics. The temperature dependence of electrolyte conductivity is a practical constraint ,many aqueous electrolytes would freeze at Mars surface temperatures without active heating, shifting some of the thermal management burden from the reactor to the electrolyte circuit rather than eliminating it. This tradeoff between thermal and electrochemical approaches depends on specific operating conditions and system architectures, and it has not been resolved by published comparative analysis at Mars-relevant detail.

Feedstock preparation shapes how efficiently either processing approach can operate. At background perchlorate concentrations of roughly half a percent by weight in bulk regolith, producing one kilogram of oxygen requires processing approximately 770 kilograms of raw regolith ,a quantity that implies substantial excavation and material handling unless the feedstock can be pre-concentrated. As discussed in Chapter 9, electrostatic pre-concentration of fine perchlorate-bearing particles has been demonstrated at laboratory scale but not validated at operational scale under Mars conditions. If pre-concentration can increase perchlorate content in the processed fraction by a factor of two to five ,as some laboratory results suggest is plausible ,the required feedstock mass drops correspondingly. Whether that concentration factor is achievable in practice at Mars depends on the specific charge properties of perchlorate-bearing particles relative to the silicate matrix under the actual humidity, temperature, and particle-size conditions at the processing site. This is measurable, but it requires a Mars surface demonstration rather than a laboratory simulation.

The oxygen production rates needed for a human mission are worth specifying clearly, because the numbers constrain what a perchlorate processing system must actually deliver. A crew of four consumes roughly four kilograms of oxygen per day for breathing under typical activity conditions. Producing that quantity from perchlorate at half-percent concentration in raw regolith requires processing approximately 3,000 kilograms of regolith per day ,roughly three metric tons ,which implies an excavation and material handling system capable of that throughput operating continuously. Producing propellant oxygen for a Mars ascent vehicle is a larger problem still: a single-stage ascent vehicle using liquid oxygen and methane propellant requires on the order of several thousand kilograms of oxygen for the ascent burn, meaning the production system must either operate for a long pre-crewed period accumulating oxygen or operate at rates substantially above life support needs during the crewed phase. For comparison, the MOXIE instrument on Perseverance demonstrated CO_2 electrolysis producing oxygen at roughly ten grams per hour at its peak performance ,a rate that, scaled to crew needs, would require a system roughly 17,000 times MOXIE's demonstrated throughput to cover breathing alone. MOXIE was a technology demonstration, not a production system, and the scaling from demonstration to production involves engineering development that is not captured by pointing at

MOXIE's peak output. This is worth stating plainly because the numbers are often cited without the scaling context that makes them meaningful.

The co-production of useful byproducts from perchlorate processing is a genuine possibility that deserves clear assessment. The solid residue from calcium perchlorate decomposition is calcium chloride, a material with real uses: it is used terrestrially as a road de-icer and as a desiccant, and its hygroscopic properties mean it could serve as a water-absorption material in a Mars surface environment. Whether having calcium chloride as a byproduct makes perchlorate oxygen production more attractive depends on whether there is actually a use for it at the scale it would be produced, which is a mission design question rather than a chemistry question. Chloride residues electrolysed further could in principle produce chlorine gas or hydrochloric acid useful in chemical synthesis, but these downstream processes add system complexity and have not been demonstrated in any integrated Mars-relevant system. They represent possibilities that warrant continued laboratory investigation without being costed into current planning assumptions.

The development trajectory for perchlorate processing as a practical Mars resource system runs through staged demonstrations of increasing scale and fidelity. A small-scale robotic demonstration on a future mission ,producing oxygen at rates sufficient for one person's breathing needs for a few days, operating autonomously before any crew is present, and generating performance data on thermal management, feedstock handling, and oxygen capture efficiency under actual Mars conditions ,would answer the questions that laboratory work cannot. Such a demonstration would be valuable regardless of whether perchlorate processing ultimately becomes a primary oxygen source or a supplementary one, because the data it generates constrains the design space for subsequent systems in ways that no Earth-based test can replicate.

Chapter 20: Water from Hygroscopic Systems

Water on Mars exists in several forms accessible at different depths and latitudes: water ice in the polar caps and in the shallow subsurface at high and mid-latitudes, water structurally bound in hydrated minerals distributed across much of the planet's ancient terrain, and water adsorbed onto mineral grain surfaces and incorporated into hygroscopic salt phases in the near-surface regolith. The first of these ,ice ,is the most mass-efficient source where it is accessible, because the water content per kilogram of processed material is high and the extraction process is conceptually simple: heat the material, capture the vapour, condense it. The second ,structurally bound water ,requires higher temperatures to liberate, typically 300 to 700 degrees Celsius for the relevant phyllosilicate and sulphate phases, and the water content per kilogram of processed material is lower. The third ,adsorbed and hygroscopic water ,is present in smaller absolute quantities per kilogram of material but can be released at lower temperatures and may be practically accessible at a wider range of latitudes and geological settings than ice.

Perchlorate's role in the hygroscopic water picture comes from its pronounced deliquescence properties. Calcium and magnesium perchlorate are among the most hygroscopic salts known, capable of absorbing water vapour from air at relative humidity levels substantially below what would cause condensation on non-hygroscopic surfaces. This property is what makes them relevant to the shallow-brine discussion in Chapter 7, but it also makes them relevant to water extraction as an engineering question independent of whether liquid brines actually form. A regolith grain coated with perchlorate that has been cycling through Martian diurnal humidity changes ,somewhat drier air during the day, slightly more humid air during the night as temperatures drop ,will accumulate adsorbed water in the form of hydrated perchlorate phases that release their water content when heated to temperatures well below those required for silicate dehydration. Laboratory measurements of water release temperatures from hydrated magnesium perchlorate show dehydration beginning around 100 degrees Celsius and proceeding through successive hydration states up to roughly 200 degrees Celsius. This is a thermal processing temperature that is achievable at reasonable energy cost, lower by several hundred degrees than what silicate mineral water extraction requires.

The water content achievable from perchlorate-bearing regolith through this low-temperature extraction route depends on how much water the perchlorate phases have accumulated. Under Mars atmospheric conditions, measured water vapour column abundances of roughly 10 to 100 precipitable micrometres vary seasonally and with latitude, with higher values in summer at mid-latitudes and near the poles during seasonal ice sublimation periods. Whether these humidity levels are sufficient to drive significant rehydration of perchlorate salts in the regolith ,or whether most of the adsorbed water is quickly lost during daytime heating before it can accumulate to useful levels ,is a question that models suggest depends sensitively on the diurnal temperature swing and regolith thermal properties at specific locations. Some modelling studies find conditions where perchlorate hydration is plausible at centimetre depths during certain seasons; others find that the conditions are marginal and that the actual water content in the near-surface perchlorate phases may be quite low under typical current Mars atmospheric conditions. The uncertainty here is real and is not resolved by existing surface measurements, because no Mars mission has directly measured the water content of perchlorate bearing material at multiple depths and times of year with the intent of testing hygroscopic accumulation rates.

This uncertainty means that hygroscopic water extraction from perchlorate-bearing regolith should be treated as a potentially useful supplementary water source whose actual yield under Mars conditions is not yet known, rather than as a reliable system whose performance can be confidently planned around. The thermochemistry works in the laboratory. The atmospheric humidity on Mars is low enough that the accumulation rate is likely modest. Whether it is sufficient to contribute meaningfully to a crew's water budget depends on numbers ,collection area, processing frequency, perchlorate concentration in collected material, diurnal hydration efficiency ,that require Mars surface measurement to constrain. A technology demonstration that accumulated perchlorate-bearing dust over a Martian winter and then measured the water released during thermal processing at controlled temperatures would directly test whether the predicted hygroscopic accumulation is occurring at the rates the models suggest. This is the kind of secondary experiment that could be accommodated on a mission primarily focused on other objectives, adding a few kilograms of demonstration hardware for a result that would substantially inform subsequent planning.

The integration of hygroscopic water extraction with the broader water supply strategy for a Mars base is a planning question whose answer depends on site. At high-latitude sites where near-surface water ice is accessible within the first metre of regolith ,confirmed at several locations by the SHARAD ground-penetrating radar instrument and by the Phoenix lander's direct exposure of ice just below the surface ,ice extraction is the preferred primary source because it delivers higher water yield per kilogram of processed material and at lower temperatures than silicate mineral extraction. In these settings, hygroscopic perchlorate extraction is a supplementary option rather than a primary one. At lower-latitude equatorial sites, where the science case for ancient habitability is often strongest but where near-surface ice is thermodynamically unstable and absent from direct measurement, the water supply options narrow. Atmospheric water capture through cold-trap condensation or sorbent cycling, regolith heating to extract structurally bound water from hydrated minerals, and hygroscopic extraction from perchlorate-bearing material become the available options rather than alternatives to ice. The relative merits of these approaches at any specific low-latitude site depend on mineral abundances, dust flux, and atmospheric humidity ,all of which vary and all of which require in-situ characterisation before a water supply architecture can be designed with confidence.

The pre-positioning concept ,deploying water collection systems robotically before crew arrival so that accumulated water is available when the crew lands ,applies to hygroscopic systems in principle but requires careful accounting of what is actually achievable over a pre-deployment period. If a collection and processing system operates autonomously for two years before crew arrival, it accumulates the water produced at whatever rate the hygroscopic system delivers under local conditions. At low collection rates, the inventory available at crew arrival might be meaningful as a reserve but not as a primary supply. At higher rates, it could contribute to initial mission water needs. The calculation depends on system-specific parameters that are not yet measured, which means pre-positioning as a strategy for hygroscopic water requires the same Mars-surface demonstration data that the broader concept needs before being relied upon in mission planning.

What the hygroscopic water concept does usefully highlight is that the water budget for a Mars base is not determined solely by the geology of a single source ,the nearest ice deposit or the nearest hydrated mineral outcrop ,but by the combined contribution of multiple sources operating at different rates and accessible under different conditions. An ice excavation system provides bulk water but requires the landing site to have accessible ice within reasonable proximity. Atmospheric capture provides small but continuous water flow wherever the atmosphere reaches. Hygroscopic regolith extraction provides access to chemically concentrated atmospheric water even at low ambient humidity. Designing a water supply that draws on more than one of these in ways appropriate to the local site conditions is more robust than optimising a single pathway, and it requires understanding all of the relevant pathways well enough to estimate their contributions at the specific site under consideration.

Chapter 21: Passive ISRU Networks

The engineering instinct when designing resource extraction systems tends toward concentration: build a central facility, bring material to it, process in bulk, distribute products outward. This model works well on Earth because the infrastructure required to support a centralised facility ,roads, power grids, maintenance supply chains ,already exists or can be built incrementally using local labour and materials. On Mars in the early phases of any surface presence, none of that supporting infrastructure exists. Every kilogram of processing equipment, every mechanism subject to wear, every component requiring periodic replacement arrives from Earth at enormous cost and competes with every other mission element for the mass allocation the launch vehicle provides. The centralised industrial model does not fail because the chemistry is wrong. It faces a scaling problem: the infrastructure needed to make it work arrives later than the infrastructure itself.

Distributed passive collection offers a different starting point, not as a replacement for all centralised processing but as a complement to it that takes advantage of what Mars already provides. The argument is straightforward in principle. Martian dust is pre-sorted by atmospheric processes, pre-charged by triboelectric interactions in the dust devil activity described earlier in this book, and deposited continuously across the surface whether or not any collection infrastructure intercepts it. A network of passive collectors ,structures with no moving parts, no continuous power requirement, and no active control during collection ,can accumulate this material at many locations simultaneously, concentrating feedstock for downstream processing without the mass penalty of excavation machinery at each node.

The analogy to solar power arrays is instructive here, though it has limits worth noting. A solar array distributed across terrain collects energy from an incident flux that the sun provides regardless of whether the panels are there. A passive dust collector distributed across terrain collects material from a transport flux that the wind and thermal convection provide regardless of whether the collectors are there. Both approaches substitute deployed area for active machinery, trading installation mass for collection area. The analogy breaks down in one important respect: solar flux at Mars is predictable, varying by a factor of roughly two between perihelion and aphelion and declining during dust storms but otherwise following a calculable seasonal cycle. Dust flux at any specific location is less predictable, depending on local wind regimes, proximity to dust source regions, and the occurrence of dust storms that are seasonal in their general timing but variable in intensity and duration from year to year. A passive dust collection network must be designed with this variability in mind, with storage capacity at each node sufficient to buffer collection periods against harvesting schedules that cannot perfectly anticipate when material will accumulate.

The mass budget argument for distributed passive collection requires honest accounting rather than optimistic estimation. A collection mesh or plate structure ,thin metal or polymer film with supporting struts and a collection hopper ,has a mass that scales roughly with the deployed area. The structural mass per unit area of a lightweight deployable collector is likely in the range of a few hundred grams per square metre for a minimal design, based on comparable terrestrial deployable structures, though Mars-specific designs optimised for the relevant wind loads, temperature cycling, and UV exposure have not been built and tested. At that mass density, one hundred square metres of collector ,roughly the footprint of a modest habitat ,adds perhaps twenty to thirty kilograms to the landed mass. Whether that mass is justified by the collection rate achieved at a specific site is a calculation that requires knowing the dust flux at that site, which requires measurement. The concept is worth pursuing in parallel with that measurement rather than waiting for perfect data before any hardware development, but it should not be planned around until collection rates under actual Mars conditions are established.

Harvesting collected material from a distributed network is an operational overhead that the concept needs to account for honestly. Manual collection by crew during periodic site visits is labour-intensive on a planet where crew time is the scarcest and most expensive resource. Automated rover-based collection is mechanically feasible in principle but adds the complexity and failure modes of an autonomous vehicle traversing the network on a regular schedule. Neither approach is free. For small pilot networks covering areas of a few thousand square metres around a base, manual collection during routine outside activities may be adequate. For expanded networks covering square kilometres, robotic collection becomes necessary, and that requirement needs to be factored into the network's mass and operational budget from the start rather than addressed as an afterthought when the network grows beyond manageable size.

The integration between distributed collection and centralised processing is where the architecture makes sense as a whole. Collection nodes accumulate raw feedstock ,dust with its perchlorate content, adsorbed water, and mineral fractions ,without power or active maintenance. That feedstock is periodically transported to a centralised processing facility where separation, thermal decomposition, water extraction, and other energy-intensive steps are performed under controlled conditions with the power infrastructure and monitoring equipment that those steps require. The logistics chain between collection and processing must be simple enough not to consume most of the efficiency gain from passive collection. If transporting feedstock from distributed nodes to a central processor costs more in crew time and rover operations than the feedstock is worth, the distributed architecture has not achieved its goal.

None of this is a reason to dismiss the concept. It is a reason to develop it in stages, starting with small instrumented pilot deployments that measure actual collection rates, characterise collected material composition, test harvesting procedures, and validate the mass efficiency claims before committing to large-area networks. A pilot deployment of ten to twenty collection nodes covering a few hundred square metres, operated alongside a primary mission with other objectives, would generate the data that current laboratory experiments and orbital dust flux estimates cannot provide. That data would either support scaling the approach or identify where the concept needs revision before further investment.

Chapter 22: Energy Systems in a Dust-Active Environment

Power on Mars is not simply an engineering trade between solar panels and nuclear reactors. It is a systems problem whose solution interacts with dust management, resource processing schedules, crew activity patterns, and the variable atmospheric conditions that characterise the Martian environment across seasons and years. Getting it wrong ,by underestimating dust degradation of solar panels, or by underestimating the mass of nuclear heat rejection systems, or by failing to account for multi-week dust storm power reductions ,has consequences that cascade across every other system the power supply supports. The operational history of Mars missions provides some useful data points, and those data points are worth examining carefully before drawing conclusions about what early human missions should use.

Mars receives solar energy at the top of its atmosphere at roughly 590 watts per square metre at perihelion, declining to about 490 watts per square metre at aphelion ,approximately 43 to 52 percent of the flux that Earth receives, reflecting both the greater average distance and the orbital eccentricity of Mars. After transmission through the atmosphere under clear conditions, surface flux is typically 500 to 600 watts per square metre at equatorial latitudes at solar noon. These numbers set the upper bound on what solar panels can collect. They are not the operational baseline. Dust reduces panel output even under nominally clear conditions, and regional or global dust storms can cut surface solar flux to a small fraction of clear-sky values for weeks at a time. During the 2018 global dust storm, solar flux at Curiosity's location in Gale Crater fell to values that, extrapolated to a solar-powered human mission, would have been operationally severe. The Opportunity rover, which was solar-powered and operating on the other side of the planet, lost enough power during that same storm that it was unable to maintain battery charge above minimum and was never heard from again.

The Opportunity failure is the most directly relevant operational data point for discussions about solar power and Mars human missions. The rover was not poorly designed. Its power system was appropriate for the mission it was designed to perform, and wind-cleaning events had extended its life far beyond its design lifetime. The 2018 storm was within the known range of Martian dust storm behaviour;global storms occur roughly every three to four Earth years, though not on a perfectly regular schedule. Designing a human mission's primary power system around a source that can fail to dust accumulation, without either active cleaning systems or backup power that sustains life-critical loads indefinitely through a multi-week storm, is a risk that the operational history suggests is not theoretical.

Small fission reactors offer the most operationally robust solution to this problem. The Kilopower project, developed by NASA and the Department of Energy and ground-tested at the Nevada National Security Site in 2018, demonstrated a reactor producing one to ten kilowatts of electrical power from a uranium-235 core using Stirling cycle converters. The system mass for a ten-kilowatt unit is approximately 1,700 kilograms, including shielding and power conversion. Higher-power variants at 40 kilowatts are under development under successor programmes. Fission reactors provide constant output independent of solar conditions, dust accumulation, time of day, or season. They require no cleaning, no battery backup for night operation, and no load shedding during dust storms. Their operational limitation is that they generate waste heat that must be rejected, and Mars's thin atmosphere makes convective cooling ineffective, so radiators ,typically thin panel structures deployed away from the reactor ,carry the thermal rejection burden. At ten kilowatts of electrical output with a conversion efficiency of roughly 25 percent, the reactor core produces about 30 kilowatts of waste heat, requiring radiator area that adds to the system's landed mass. This is a real cost, but it is a calculable and manageable one, and the mass penalty of radiators is not a fundamental barrier to nuclear power at the scales needed for early crewed surface operations.

Solar power is not without a role in a Mars surface power architecture, but the role it plays most efficiently is as a supplement to nuclear baseload rather than as the primary source with battery backup against dust storms. During clear conditions, solar panels can provide peak power above the nuclear baseload for energy-intensive activities like propellant production or high-power research operations, while the nuclear system handles continuous baseline loads ,life support, communications, thermal management, and the control systems that run continuously regardless of crew activity. This hybrid architecture accepts the mass of both systems in exchange for operational resilience: solar augmentation during favourable periods, nuclear continuity during storms and nights. The redundancy it provides is not elegance for its own sake. It is the engineering response to operating on a planet where a single severe dust storm can last long enough to be fatal to a solar-only system.

Battery storage on Mars is subject to the same mass constraints that govern every landed system, and the energy density of current battery technology limits how much storage is practical. Lithium-ion cells at roughly 200 to 250 watt-hours per kilogram ,a reasonable estimate for near-term technology ,store energy at a mass cost that becomes prohibitive for multi-day gap-filling at the power levels human missions require. Supplying ten kilowatts of baseload for fourteen Martian nights of roughly twelve hours each ,roughly the energy a crew would need through two Earth weeks without any solar input ,requires on the order of 2,500 to 3,000 kilowatt-hours of storage, which translates to ten to fifteen metric tons of battery mass at current energy densities. That is not a practical battery backup system for a crewed Mars mission. It is an argument for nuclear baseload, which produces the same continuous power for the same mission duration at a fraction of the mass, rather than for larger batteries.

The intersection between energy systems and dust-driven resource collection deserves a specific and honest assessment, because the idea that triboelectric energy harvesting from Martian dust could constitute a useful power source occasionally appears in discussions of Mars resource utilisation. The physics of triboelectric charge generation in dust devils is real, and the principle that a conductive structure in an electric field experiences a force on charge carriers that could in principle drive a current has laboratory-scale demonstrations. What has not been demonstrated is any system that harvests this effect at useful power levels under Mars conditions. The charge densities in Martian dust devils, while higher than terrestrial equivalents due to the lower atmospheric pressure, are not large in absolute terms. Estimating the power harvestable from a collector of given area in a dust devil of given size and field strength produces numbers in the milliwatt to watt range for realistic collector configurations ,useful as a curiosity and worth investigating through small flight demonstrations, but not a power source that meaningfully supplements a multi-kilowatt life support system. It should be characterised as a speculative research direction, not a planned power contribution.

Waste heat integration is an area where careful system design produces genuine efficiency gains without adding mass. A nuclear reactor or a fuel cell operating aboard a Mars habitat rejects heat that, if simply dumped to the environment, must be replaced by active heating of living and working spaces where temperatures without heating would be lethally cold. Routing waste heat through habitat thermal management before final rejection to radiators recovers energy that would otherwise be lost twice ,once in the conversion inefficiency of the power system, and once in the heating system that must replace it. Designing these thermal circuits into the habitat from the start, rather than retrofitting heat recovery into a system originally designed without it, is the straightforward engineering approach. It requires system-level thinking at the design stage, which is less common than component-level optimization but produces better overall performance.

Chapter 23: Preservation-Aware Exploration Systems

The equipment and operational strategies used in Mars astrobiology missions reflect the assumptions embedded in the science programme that designed them. Current rovers drill to roughly five to six centimetres, reflecting an assumption that surface and near-surface samples contain interpretable chemistry. Instrument suites are calibrated for the concentration ranges expected from geologically interesting surface materials, reflecting assumptions about what those materials contain. Landing site selection prioritises ancient aqueous environments, reflecting the assumption that where water was present, biosignatures are best sought. These assumptions are not unreasonable given what was known when each mission was designed. The preservation gradient framework developed in this book argues that at least one of them ,the assumption that near-surface samples at well-chosen ancient sites contain adequately preserved chemistry ,needs to be revisited in light of what is now understood about ongoing atmospheric oxidation.

Revising that assumption has practical consequences for hardware, operational strategy, and site selection that are worth working through specifically. The most direct hardware consequence is drill depth. If the oxidation zone at a site with substantial surface exposure age extends to tens of centimetres or deeper in permeable material, samples collected from five to six centimetres are almost certainly within that zone at most scientifically interesting ancient sites. Accessing material below the oxidation transition ,where chemistry grades from modified toward original ,requires a drill capable of reaching at least several tens of centimetres and ideally a metre or more in coherent rock. No Mars rover currently carries this capability. The engineering path toward it exists: lightweight pneumatic and percussion drill systems capable of metre-scale penetration in competent rock have been developed in terrestrial contexts, and adapting them for Mars involves solving real but not fundamental engineering problems around mass, power consumption, core handling, and operation in low-pressure cold environments. This adaptation requires development investment and flight qualification that takes years. The time to begin that development is before the next generation of astrobiology missions is in its final design phase, not after.

In parallel with deeper drilling, the ability to measure oxidation state in situ at each depth interval ,rather than only analysing collected cores ,would enable adaptive sampling strategies where drilling continues deeper only at locations where the depth profile warrants it. An instrument capable of measuring the ratio of ferric to ferrous iron in rock at the drill face, or measuring perchlorate concentration in gases evolved from a small heated sample at each depth step, would provide real-time information about whether the sample at that depth is within the oxidation zone or below it. This kind of downhole measurement is standard practice in terrestrial borehole geochemistry and in oil and gas exploration; adapting the relevant sensing principles to a small, low-power Mars drill system is a specific and achievable engineering objective. It would make depth profiling campaigns far more informative than sampling at fixed predetermined depths, because it would allow the drill to find the transition zone rather than hoping a fixed depth bracket crosses it.

Landing site selection that incorporates preservation considerations alongside ancient habitability requires adding quantitative data layers to the site evaluation process currently used by mission selection committees. The relevant datasets ,Context Camera dust devil activity maps spanning multiple Martian years, surface age estimates from crater size-frequency analysis, mineralogical maps from the Compact Reconnaissance Imaging Spectrometer for Mars ,already exist in the planetary data archives. They have not been systematically combined into preservation quality metrics and applied to candidate landing site comparisons. Doing so requires combining the datasets in a geographic information framework, defining quantitative preservation indices that capture the combined effect of dust activity, exposure age, and lithological permeability, and making those indices available to site selection teams alongside the habitability indices that currently dominate discussions. The analysis is technically straightforward, though the specific index formulations involve choices about how to weight different variables that require validation against ground truth ,validation that will come from the same multi-site chemical comparisons that the framework predicts should show preservation differences.

Time-series chemistry measurements at established landing sites could test whether surface oxidant levels respond to seasonal dust activity cycles, providing direct evidence for or against the active production hypothesis without requiring a new landing site. Curiosity has now operated in Gale Crater for more than six Earth years across multiple Martian year cycles. If rock surface measurements or drill-sample analyses from early in the mission are compared systematically with measurements from later years at the same or comparable locations, any seasonal or year-to-year chemistry variation would be detectable against the instrumental baseline. The challenge is that this kind of repeat measurement at identical locations was not originally designed into mission operations, and instrument drift over years of operation complicates direct comparison of early and late readings. It is, however, a data analysis question that could be investigated with the existing Curiosity dataset if the analysis were prioritised by the science team. The result ,whether surface chemistry at Gale Crater shows any seasonal pattern correlated with dust devil activity levels ,would be one of the few direct tests of active production that the existing surface mission data could provide.

The coordination between orbital characterisation and surface investigation is already standard practice in Mars mission planning in the sense that orbital data informs where rovers go. What is not yet standard is using orbital data to characterise the modern chemical processing history of candidate sites ,their dust activity levels, their exposure ages, their rock type proxies from spectroscopy ,in a way that feeds into preservation quality estimates rather than only into habitability assessments. The information needed to do this is largely in hand. The dust activity characterisation from Context Camera analysis is documented and available. The surface age maps from crater counting are published. The mineralogical maps from the Compact Reconnaissance Imaging Spectrometer are in the Planetary Data System. Assembling them into a preservation-quality framework for landing site evaluation is a science planning task, not a new data collection campaign, and it is one that could be carried out in advance of the next major astrobiology mission selection cycle if the effort were made.

Sample return is the most powerful endpoint for preservation-aware exploration because terrestrial laboratory analysis can measure everything that orbiter and rover instruments cannot: perchlorate isotopic composition, full molecular structure of complex organics, micrometer-scale mineral texture, trace element distributions at parts-per-trillion sensitivity. But returned samples only answer questions about the locations and depths from which they were collected. If Perseverance's cached samples come predominantly from surface and near-surface exposures at a moderate-activity site, they will reflect the chemistry of the oxidation zone as much as any hypothetical ancient biological chemistry beneath it. Designing the caching strategy to deliberately include depth profiles from freshly drilled cores, where each core section represents a different depth in a potential oxidation gradient, would make the returned sample set more interpretable from a preservation standpoint. This is a sample selection decision that can still be influenced by how the science community frames the question, and framing it explicitly ,cached samples should include depth series, not only lateral diversity ,is the straightforward advocacy that the preservation framework motivates.

Chapter 24: Site Selection for Settlement

The first human settlement on Mars will be constrained by factors that robotic mission landing site selection rarely needs to prioritise: reliable access to water in extractable quantities, terrain accessible enough for construction activities beyond a few square metres, a power supply capable of sustaining life-critical systems through worst-case seasonal and storm conditions, and sufficient flat ground for the landing systems that will bring people and cargo over a period of years. These are not the criteria that have driven the selection of Jezero Crater or Gale Crater as science targets, and the divergence between what makes an excellent astrobiology site and what makes a practical settlement location is one of the more consequential tensions in long-term Mars planning.

Water is the clearest example of that tension. The science programme has rightly concentrated on equatorial and low-latitude sites where ancient aqueous environments are well-preserved in mineralogy and stratigraphy and where solar power is most reliable and diurnal temperature swings are most moderate. Near-surface water ice at these latitudes is not accessible ,ice is thermodynamically unstable within the first few metres of the surface at latitudes below roughly 30 to 40 degrees, sublimating into the atmosphere rather than persisting. The same thermal properties that make low-latitude sites comfortable by Martian standards also make them poor locations for finding the water ice that is the most mass-efficient water source for a settlement. The confirmed near-surface ice ,detected directly by Phoenix at around 68 degrees north latitude, inferred from neutron spectrometer data across mid-to-high latitudes, and mapped by SHARAD ground-penetrating radar at multiple locations ,is at latitudes where average temperatures are lower, diurnal temperature swings are larger, and the solar power environment in winter is poor enough that nuclear power becomes effectively mandatory rather than optional.

The mid-latitude band between roughly 30 and 50 degrees in either hemisphere represents the practical compromise space where several requirements can be partially satisfied simultaneously. Near-surface ice may exist at accessible depths ,a few metres rather than centimetres ,based on thermal models and radar sounding, though direct confirmation through drilling is limited to the Phoenix landing site and a handful of other locations. Solar power at these latitudes is usable through most of the year with reduced winter output. Terrain in much of this latitude band is geologically ancient and relatively flat compared to the heavily cratered highlands further south or the steep volcanic provinces near the equator. The scientific interest for ancient habitability is lower than at confirmed lake sites like Jezero, but it is not absent ,phyllosilicate minerals indicating aqueous alteration are distributed across ancient terrains at these latitudes, and the preservation quality of ancient materials in low-activity portions of this terrain may actually be better than at the more celebrated equatorial science sites, for the reasons developed in earlier chapters.

Dust activity is one of the variables that varies systematically with latitude and terrain type in ways that matter for both operational burden and long-term resource availability. High-activity zones identified from Context Camera surveys tend to cluster in regions with low thermal inertia surface materials ,fine dust and sand that heat rapidly ,at latitudes between roughly 10 and 40 degrees where surface heating drives strong boundary layer convection. A settlement located in the heart of one of these high-activity zones would contend with continuous dust accumulation on solar panels, elevated perchlorate in ambient dust posing a crew health management challenge, and equipment wear from abrasive dust entering mechanisms. The same high-activity location would theoretically provide a higher supply rate of perchlorate-bearing fine material for electrostatic collection, if that approach is validated as useful. Whether the resource advantage of high dust activity outweighs the operational penalty is not currently answerable with confidence, because neither the collection efficiency of electrostatic systems under actual Mars conditions nor the specific maintenance cost of dust-intensive operations on human-rated equipment has been measured. Settling this question is part of what precursor robotic missions to candidate settlement sites should accomplish before human crews are committed.

Elevation affects landing performance in ways that directly limit where large payloads can be delivered. Mars entry, descent, and landing systems using aerodynamic deceleration ,parachutes, inflatable decelerators, or lifting entry vehicles ,require atmospheric density sufficient to generate meaningful drag force before terminal velocity is reached. At the highest Martian elevations, particularly the Tharsis plateau and the flanks of the major shield volcanoes, the atmosphere is thin enough that parachutes cannot alone slow a large payload to safe terminal velocity before ground impact. The lower the elevation, the denser the atmosphere and the more aerodynamic deceleration assists the landing system. The Hellas impact basin, reaching roughly eight kilometres below the mean Martian radius, has the densest near-surface atmosphere on the planet and from a landing physics standpoint is one of the most accessible locations for heavy payloads. Its geological setting ,ancient heavily cratered basin floor with limited mineralogical evidence for recent aqueous activity ,makes it scientifically less compelling for astrobiology, but for a settlement whose priority is sustainable operations rather than ancient habitability investigation, that tradeoff may be acceptable. No decision has been made about Hellas as a settlement site and significant site characterisation at ground level would be needed before any serious planning could begin, but it illustrates that the criteria relevant to settlement diverge from those relevant to science in ways that the community should be explicit about rather than assuming overlap.

The dust activity framework developed in this book adds a variable to settlement site selection that conventional analyses do not include:

the degree to which ongoing atmospheric chemistry has modified surface materials over geological time, and the consequences of that for both resource extraction and any scientific investigation conducted from the settlement. A settlement located in a region of consistently low dust activity sits in terrain where ancient surface materials are less oxidised, where perchlorate concentrations in the regolith may be somewhat lower than in high-activity regions, and where the operational dust burden on equipment and crew health management is reduced. These are real advantages from both a science and an engineering standpoint, and they are quantifiable ,not yet quantified with the precision that formal site ranking requires, but quantifiable in principle from the orbital datasets already in hand. Incorporating them into settlement site evaluation is not a departure from engineering rigour. It is an extension of the same rigour that already applies to power, terrain, and landing system constraints.

The irreversibility of settlement site selection is the overriding reason why the investment in thorough precursor characterisation is justified regardless of the cost. A robotic mission that demonstrates water extraction rates, measures dust flux and perchlorate concentrations, validates solar panel degradation rates and cleaning approaches, characterises regolith mechanical properties for construction, and tests landing system performance at a candidate site before human crews are committed is not a luxury. It is the minimum responsible precursor for a decision whose consequences cannot be undone once infrastructure is built and people are present. The history of terrestrial settlement includes many examples of sites chosen on incomplete information that proved problematic only after the commitment was made. Mars offers no correction mechanism comparable to the ship that can carry people home when a terrestrial settlement fails. Getting the site selection right the first time is not just a preference. It is a requirement.

Chapter 25: The Mars Industrial Model

Industry on Mars will begin small and stay small for longer than optimistic timelines usually acknowledge. The first crews will number in the single digits. Their infrastructure will have been designed to the minimum mass needed to keep them alive, working, and capable of returning home if required. The industrial activity of that initial phase ,if it can be called industrial at all ,will consist largely of maintaining imported systems, conducting limited resource extraction to supplement brought supplies, and generating the operational experience that informs what the next mission should bring. This is not a failure of ambition. It is the honest consequence of the physics and economics of interplanetary logistics, which impose constraints on early Mars operations that no amount of technological optimism changes.

The characteristic that distinguishes Mars industrial development from every terrestrial precedent is the complete absence of supply chain depth. On Earth, even the most remote industrial operation can eventually receive a replacement part, a specialist technician, or a bulk material shipment within weeks or months. On Mars, the nearest supply depot is at minimum six months away by the fastest trajectory, and the launch window that allows the journey opens only once every roughly twenty-six months. A component that fails without a spare on hand is simply failed. A material that runs out cannot be replenished until the next mission arrives, if the next mission brings it. This constraint shapes every decision about what to manufacture, what to repair, what to stockpile, and what to do without in a way that has no close terrestrial analogy. The Antarctic overwintering experience comes closest ,small crews isolated for months with finite supplies ,but even Antarctic stations receive resupply by air, and they are never more than a few days from evacuation if something goes seriously wrong.

The practical response to this constraint is not to try to replicate terrestrial industrial capability at small scale, which is both mass-inefficient and functionally inadequate at the throughput rates an early settlement needs. It is to design systems around the materials and processes that Mars provides locally, and to design those systems to be maintainable with tools, skills, and spare materials that the crew already has. Standardisation of fasteners, seals, electrical connectors, and structural profiles across different systems reduces the unique part inventory that must be stocked. Designing components to be repairable rather than disposable ,with accessible fastening, interpretable failure modes, and materials that can be worked with basic tools ,reduces dependence on exact replacement parts. These design philosophies are in tension with mass-minimisation approaches that shave every component to its minimum weight, because minimum-mass components are often minimum-repairability components as well. For Earth missions where mass is the binding constraint and operational lifetime is months, minimum-mass design is appropriate. For a settlement where operational lifetime is years and resupply is infrequent, repairability must be weighted alongside mass in the design trade.

The materials available for local manufacture on Mars are constrained but not negligible. Regolith is abundant and chemically diverse ,basaltic silicates, iron oxides, sulphates, perchlorates, and minor quantities of other phases are present in quantities limited only by excavation and processing capacity. Basalt can be melted and cast into structural shapes, a process that requires high temperatures but no imported feedstock. Compacted regolith can serve as radiation shielding and thermal mass, reducing the volume of pressurised habitat that must be imported. Iron can in principle be reduced from iron oxide minerals using hydrogen or carbon monoxide as reductants, producing usable metal at the cost of the reducing agent and the energy to drive the reduction. The temperatures required for iron smelting

,above 1,500 degrees Celsius for iron oxide reduction to proceed rapidly ,are achievable with concentrated solar or with resistance heating from a nuclear power source, but they require a reactor design and containment system that go well beyond early-mission capability. The realistic trajectory for local metal production begins not with ore smelting but with repair welding and casting of small components from collected iron-bearing material, building toward more ambitious metallurgy as power and infrastructure develop over successive mission cycles.

Additive manufacturing ,3D printing in its various forms ,is frequently cited as a transformative capability for Mars because it promises to produce arbitrary part geometries from bulk material feedstock, potentially replacing the need to stock replacement parts for every conceivable failure mode. The reality is more constrained. Fused deposition modelling with polymer filament can produce plastic parts of moderate structural complexity from feedstock that is itself manufactured on Earth from petrochemicals, not from Martian resources. Printing structural metal components using powder bed fusion or directed energy deposition requires metal powder feedstock and laser or electron beam energy sources that together constitute a mass-intensive system producing parts that, for structural applications, often do not match the fatigue and fracture properties of wrought or cast equivalents. For electronics, sensors, and precision optical components, additive manufacturing provides essentially nothing;these require manufacturing processes with multiple steps and specialised precursor materials that go far beyond what any near-term Mars system could support. Additive manufacturing is genuinely useful for producing non-structural replacement parts with complex geometry where the performance difference from conventionally manufactured equivalents does not matter ,brackets, housings, ergonomic tools, non-critical fluid connectors. It is not a general solution to the spare parts problem, and presenting it as one gives a false impression of how self-sufficient an early settlement can be.

The energy intensity of small-scale manufacturing is consistently higher than large-scale equivalents because economies of scale in energy efficiency work against small operations. A small batch furnace processing a few kilograms of material per cycle consumes more energy per kilogram processed than a large continuous furnace at industrial scale, because the energy required to heat the furnace structure and maintain temperature represents a larger fraction of total energy input when throughput is small. This is a fixed thermodynamic reality, not an engineering deficiency that can be designed away. It means that on Mars, where power is the limiting resource in early phases, manufacturing processes must be chosen to minimise energy per unit output rather than maximise throughput, and processes that can be run using waste heat from power generation or from other industrial steps are preferable to processes requiring dedicated high-temperature energy input.

Quality control under Mars conditions is more demanding than quality control on Earth, for a reason that the supply chain constraint makes immediately obvious. On Earth, a component that fails inspection can be returned to the supplier, remanufactured, or replaced from stock. On Mars, a component produced locally with a flaw that is not caught before installation may fail in service in a life-critical system with no replacement available. The consequence of poor quality control is not an inconvenient delay. It may be a fatality. This reality argues for building quality measurement capability into manufacturing systems from the beginning ,not as an optional upgrade but as a primary requirement ,even though measurement instrumentation adds mass and complexity. The alternative is producing components of unknown quality and hoping they work, which is not an engineering posture appropriate to the environment.

The industrial model that emerges from these constraints is not the model of terrestrial factory production transplanted to another planet. It is something more deliberate and more conservative: systems designed for longevity over performance, repairability over minimum mass, redundancy over optimisation, and local material compatibility over imported precision. This is not primitive technology. It is appropriate technology ,a term that in this context means technology matched to the actual constraints and resources of its operating environment rather than technology selected by analogy to how things are done somewhere with different constraints. The Mars environment is different from Earth in ways that matter for industrial design. Acknowledging those differences and designing accordingly is the basis for a sustainable Mars industrial capability, however modest that capability must be in the early phases of what will necessarily be a long process of incremental development.

Recycling is not an environmental ethic on Mars. It is an economic necessity of the first order. Every kilogram of material that arrived at the surface represents a launch cost that, at even the most optimistic projections for heavy-lift reusable vehicles, amounts to thousands of dollars. Discarding a worn component and waiting for a replacement wastes that original investment and adds the cost and delay of replacement. Recovering the material from a failed component and reusing it in a new application recovers a fraction of that investment. The energy cost of recycling ,melting, reprocessing, reforming ,is almost always lower than the energy equivalent of the launch cost it avoids. The economic case for aggressive material recovery is compelling at current and near-future launch costs, and it becomes less compelling only if launch costs fall to a level where resupply becomes cheap relative to the cost of maintaining recycling infrastructure. That point is not yet in sight, and planning for the first several generations of Mars surface operations should treat material recovery as a baseline design requirement rather than a supplementary consideration.

The long arc of Mars industrial development, if human presence on the planet sustains itself and grows, runs from the first small crews maintaining imported systems toward progressively more self-sufficient communities producing an expanding fraction of their material needs locally. Each stage of that progression enables the next ,a machine shop produces tooling needed for more advanced fabrication, which produces components needed for more capable energy systems, which powers more ambitious material processing. The pace of progression depends on how many people are present, how much power is available, and how successfully each generation of infrastructure supports the next. What the trajectory does not look like is a sudden transition from dependency to self-sufficiency. It is gradual, cumulative, and dependent on continuous investment in the infrastructure that makes each next step possible. Recognising this is not pessimism. It is the realistic basis for planning a programme of Mars industrial development that succeeds because it accounts for the constraints it actually faces rather than the constraints it wishes were true.

The industrial model described in this chapter is a framework derived from first principles and operational analogy. It identifies the constraints that will govern early Mars manufacturing and the design philosophies those constraints favour, but it stops short of specifying the hardware. That is the work of the chapters that follow. Part Three moves from the what to the how ,from the argument that Mars's ongoing electrochemistry is both a resource to be used and an engineering context to be designed within, to the specific systems that have been built, tested, and in some cases operated on the Martian surface that begin to show what closing those loops actually looks like in hardware. The transition is deliberate. The science and systems thinking of the preceding chapters are not preamble to the engineering. They are the basis on which the engineering decisions can be made rationally rather than by analogy to terrestrial practice in a setting where terrestrial practice does not apply.

Part Three: Engineering the Electrochemical Mars

The first two parts of this book established that Mars is not a passive geological museum but an active chemical system, one that manufactures oxidants through triboelectric discharge in dust devils, concentrates hygroscopic salts through atmospheric transport, and imposes a spatial structure on its surface chemistry that directly affects both what can be found there and what can be used. The implications of that framework reach further than exploration strategy. They reach into hardware design, mission architecture, and the sequence of decisions that must be made before the first human crews can sustain themselves on the surface.

This section approaches those engineering questions directly. It distinguishes between hardware that has been flight-validated on Mars, systems that have been demonstrated under simulated Mars conditions on Earth, physical mechanisms that are established in the laboratory but not yet tested in a Mars-relevant integrated context, and architectural concepts that are grounded in demonstrated physics but have not yet been built in any form. These are different things, and treating them as equivalent is the shortest route to proposals that look compelling on paper and fail in implementation.

The central claim that runs through what follows is narrower and more defensible than it might first appear. Mars has been concentrating and processing certain materials through its atmospheric chemistry for billions of years. The question is not whether that processing has occurred ,the observational evidence is clear that it has ,but whether human engineering systems can be designed to interface with it in ways that reduce mission mass, improve resource access, and provide capabilities that purely Earth-sourced hardware cannot match. That question has a serious engineering answer, and it is attempted here.

Chapter 26: What the Hardware Has Confirmed

Any engineering argument must begin with what is actually known. Three developments have changed the foundation of Mars resource engineering since 2021, and understanding their significance precisely ,neither overstating nor understating what they proved ,is prerequisite to everything that follows.

The triboelectric mechanism: from inference to direct observation

In October 2024, the SuperCam instrument aboard Perseverance recorded something that planetary scientists had been predicting for decades. As a dust devil passed directly over the rover, the instrument's microphone detected three sharp acoustic events between the sounds of the vortex's leading and trailing walls. Those events were electrical discharges ,the triboelectric crackling that atmospheric chemists had attributed to the mechanism described throughout this book's earlier chapters, now captured on audio in real time on the surface of Mars. By the completion of the analysis, the science team had logged fifty-five distinct electrical events associated with dust devils and storm fronts, establishing that Martian atmospheric conditions routinely generate charge separation sufficient to produce actual breakdown events.

The engineering significance of this observation is specific. It confirms that the charge accumulation in Martian dust vortices reaches the threshold required for arc formation under Mars surface pressure conditions. It confirms that the phenomenon is not marginal or rare but occurs across a range of atmospheric conditions and vortex types. And it confirms that the physics underlying the triboelectric chemistry framework ,the framework that this book uses to explain perchlorate distribution, oxidant production, and surface alteration patterns ,operates at scales and intensities that have real consequences for surface materials and, by extension, for any hardware placed on the surface.

What this observation does not confirm is the specific rate of perchlorate production at any given location, the enrichment factors achievable through electrostatic collection in actual Mars surface conditions, or the spatial correlation between discharge intensity and measured surface chemistry. Those remain questions for instruments and missions that have not yet flown. The distinction matters because the confirmed observation is the foundation for an engineering approach, not the validation of a specific system design.

One qualification on the production confirmation is worth stating explicitly for the engineering reader. The triboelectric mechanism produces perchlorate, and that production is now observed behaviour rather than inference. What recent laboratory and modelling work also establishes is that galactic cosmic rays decompose perchlorate at the surface through radiolysis, producing chlorate, hypochlorite, and molecular oxygen as breakdown products. The surface concentration of roughly half a percent by weight measured at Phoenix and Curiosity therefore reflects a dynamic equilibrium between production and destruction rather than a purely accumulating inventory. From a resource engineering standpoint this is not a problem ,a steady-state concentration is more reliable to plan around than one still climbing toward an unknown maximum ,but it does affect how the collection system should be designed. Radiolytic destruction is most intense at the very surface, where cosmic ray flux is highest, and decreases with the depth of overburden. This means the feedstock most worth targeting may be slightly subsurface material, a few centimetres below the immediate surface layer where the destruction rate is most active, rather than the dust deposited at the air-regolith interface. The Phase 1 demonstration should include depth-stratified sampling alongside surface collection to characterise whether this gradient is measurable and whether it is large enough to influence collector design.

Solid oxide electrolysis: from demonstration to established technology

The MOXIE experiment completed its mission in August 2023 after sixteen oxygen production runs spanning a full Martian year. It produced one hundred and twenty-two grams of oxygen from the Martian atmosphere, achieving a peak production rate of twelve grams per hour at ninety-eight percent purity ,twice the performance target its designers had set. It operated successfully at different times of day and across different seasons, demonstrating that the process is not limited to favorable atmospheric conditions. When the program concluded, NASA described solid oxide electrolysis of Martian atmospheric carbon dioxide as a technology at the highest readiness level, validated in its intended operating environment.

The scaling arithmetic from MOXIE to a human-mission system is engineering rather than research. A unit two hundred times the size of MOXIE, paired with a power plant in the forty-kilowatt class, would produce oxygen at approximately two kilograms per hour continuously. Operating autonomously for one Earth year before crew arrival, such a system would accumulate roughly seventeen tonnes of liquid oxygen, sufficient to fuel a Mars ascent vehicle for a six-person crew. That mass of oxygen, if brought from Earth rather than produced on Mars, would require thousands of additional tonnes of propellant across the launch and transit sequence. The economic argument for in-situ oxygen production is no longer open; the question is what to combine it with.

Subsurface ice access: demonstrated capability, not yet flight heritage

Honeybee Robotics' RedWater system combines coiled-tube drilling, adapted from terrestrial oil and gas practice, with Rodriguez Well technology developed for Antarctic ice access. The concept has been tested under simulated Mars conditions at increasing fidelity. The system is designed to drill through up to twenty meters of unconsolidated overburden, reach subsurface ice, establish a melt cavity, and pump liquid water to the surface without continuous excavation. At mid-latitude sites where glacial ice deposits have been mapped by orbital radar sounders, a mature RedWater system operating for one Mars year could extract tens of tonnes of water from a resource that requires no propellant processing and no atmospheric chemistry.

The important qualification is that RedWater has not flown to Mars. Its demonstrated capability exists in Earth-based test environments. The transition from terrestrial demonstration to Mars flight involves engineering challenges ,regolith composition variability, thermal behavior at Mars temperatures, autonomous fault response without crew ,that are not trivial. This places the system at a technology readiness level that warrants serious engineering investment rather than immediate mission commitment. The distinction matters because early human missions will depend on water, and systems that have not operated autonomously under actual Mars surface conditions carry risk that systems with flight heritage do not.

The hardware gap that shapes everything else

The Curiosity and Perseverance drills both reach approximately seven centimeters into consolidated rock. The HP3 mole on InSight was designed to reach five meters through regolith but was unable to generate sufficient friction in the cohesive soil at its landing site and ceased operations before reaching fifty centimeters. Every in-situ compositional measurement performed on Mars rock to date has sampled the uppermost few centimeters of material ,precisely the zone that this book's earlier chapters argue is most affected by ongoing oxidative alteration.

The consequence is direct: any mission architecture that does not include meter-scale subsurface access is structurally incapable of testing the oxidation gradient hypothesis, validating preservation models at depth, or demonstrating access to resources known to exist below the surface alteration zone. This is not a scientific limitation. It is a hardware limitation, and it applies equally to human and robotic missions.

JPL's Mars Deep Drill program has developed a wireline percussive system targeting one hundred meters of penetration in a payload under one hundred kilograms drawing one hundred watts. Hardware prototypes have been tested in relevant terrain. The system has not flown. NASA's 2024 Mars exploration planning documents explicitly identify the need for affordable deep drilling access as a priority capability gap. The FY2026 budget request includes three hundred and fifty million dollars for a new Mars Technology program intended to accelerate high-priority technologies, with fission surface power and deep access both cited in context. The institutional recognition that this gap exists is now formal. What is needed is the mission architecture to close it.

Technology readiness: a summary

The following table summarizes the readiness status of the key components discussed in this section. The assessments are intended to be conservative rather than promotional. Systems with flight heritage on Mars are classified at the highest levels. Systems with Earth-based demonstration under simulated conditions are classified at middle levels. Architectural concepts grounded in demonstrated physics but not yet built as integrated systems are classified at low levels. This hierarchy is the baseline against which all subsequent engineering arguments must be held.

Component	TRL	Basis
CO_2 solid oxide electrolysis (MOXIE)	9	16 successful runs on Mars across full Martian year; 12 g/hr at 98% purity
Triboelectric discharge on Mars	9	55 electrical events recorded by Perseverance SuperCam, Oct 2024
Perchlorate thermal decomposition	5–6	Established reaction chemistry; demonstrated in Earth lab with Mars simulants
Electrostatic dust/mineral separation	4–5	Operational in terrestrial industry; adapted lab tests with Mars simulants at low pressure

Hygroscopic water recovery from residues	3–4	Thermodynamics well characterized; limited Mars-analog lab testing
Coiled-tube subsurface ice drilling (RedWater)	5–6	Field-tested under simulated Mars conditions on Earth
Meter-scale rock drilling on Mars	3–4	JPL Mars Deep Drill prototyped; field-tested in analog terrain
Integrated electrochemical ISRU stack	2–3	Individual components at various TRL levels; integration architecture is conceptual
40–100 kW class fission surface power	4–5	Phase 1 designs completed; Phase 2 solicitation

active; lunar demo target early 2030s

Chapter 27: A Proposed Integration Architecture

The following architecture represents a proposed integration of independently validated subsystems. While each component is grounded in demonstrated physics or flight hardware, the system as a whole has not been validated in an integrated Mars-relevant environment. It is presented as a coherent engineering direction rather than a deployable design, and the distinction between those two things is material.

The case for integration is not primarily scientific. Every kilogram delivered to the Martian surface costs the same regardless of what that kilogram is doing. A system designed as a collection of independent subsystems, each with its own power supply, control electronics, structural support, and thermal management, weighs more than a system designed from the start as a connected material flow. The question integration is trying to answer is whether the mass savings from shared infrastructure justify the engineering complexity of connecting components that were originally developed separately. For the specific combination of capabilities described here, the argument is that they do ,but the claim is architectural, not operational.

The material flow

The starting point is ambient dust. Mars delivers electrostatically pre-charged, size-sorted fine particles to any surface exposed to the atmosphere, with delivery rates that are predictable in their seasonal structure from orbital dust devil observations. Rather than treating this as a contamination problem to be excluded, the architecture treats it as a feedstock stream to be intercepted.

A passive collector array ,parallel conductive mesh panels maintained at alternating electric potentials ,creates field gradients that deflect incoming charged particles toward collection surfaces. The mesh geometry determines which particle sizes are preferentially retained. Terrestrial triboelectrostatic belt separators operating on the same physical principle process fine mineral powders at industrial scale with power demands of one to four kilowatt-hours per tonne of feed material. On Mars, the lower atmospheric pressure means ambient particles arrive more highly charged than in terrestrial applications, because the sparse atmosphere provides fewer ions to dissipate accumulated charge. Whether this translates into meaningfully higher collection efficiency or preferential enrichment of perchlorate-bearing fines relative to silicates is a question that has not been answered by flight data. Order-of-magnitude estimates based on laboratory analog work suggest that enrichment factors of two to five times bulk regolith perchlorate concentration may be achievable, but these estimates carry wide uncertainty and must be treated as targets for a demonstration mission rather than assumptions in a production design.

The feedstock concentration assumed in the collection architecture also needs to be understood as a steady-state value rather than a continuously growing one. Galactic cosmic rays decompose perchlorate in the near-surface regolith through radiolysis, meaning the measured bulk concentration of roughly half a percent by weight represents a dynamic balance between triboelectric production and radiation-driven destruction. This has one specific design implication for the collector array: the optimal collection target may be slightly subsurface material rather than ambient surface dust, because the very top layer of the regolith sees the highest cosmic ray flux and therefore the most active radiolytic destruction. Whether the gradient between the surface and a few centimetres below it is large enough to matter for collection efficiency is a question the Phase 1 depth-stratified sampling will answer. The steady-state nature of the concentration is otherwise an engineering advantage ,it means the feedstock supply rate is stable and predictable rather than dependent on accumulation time since the last disturbance.

Collected material passes to a two-stage electrostatic separator. The first stage uses inter-particle contact in a vibrational feeder to enhance charge separation between mineral species, then drops the material between parallel electrodes at several kilovolts. Different mineral phases accumulate different charge polarities; perchlorate-bearing particles, which are hygroscopic and have characteristically different surface electrical properties from dry silicates, are expected to deflect preferentially toward one electrode. The second stage processes the enriched fraction again for additional concentration. The power requirement for this unit is modest relative to primary operations ,on the order of one to two hundred watts during processing cycles ,and it produces no chemical waste stream, only separated mineral fractions going to different downstream processes. The silicate fraction separated at this stage ,the perchlorate-depleted mineral material that deflects toward the opposite electrode ,is not a waste stream. It is construction feedstock, and the fact that the electrostatic separation has removed most of its perchlorate content matters for what can be done with it.

The primary candidate for producing structural materials from Martian regolith is sintering: heating the material to between roughly 1,000 and 1,150 degrees Celsius until mineral grains fuse into load-bearing blocks. Laboratory work with Martian regolith simulants has demonstrated compressive strengths of 45 to 125 megapascals in sintered products depending on mineralogy and process temperature ,strengths that substantially exceed what structural applications under Mars's reduced gravity require. Microwave sintering has shown particular energy efficiency advantages over conventional thermal sintering, with energy consumption estimates on the order of 200 megajoules per tonne of processed material. The structural case for sintered regolith bricks as a primary construction material for early Mars infrastructure is credible and well-supported in the materials literature.

The problem that this literature has only recently begun to confront is perchlorate. Standard Martian regolith simulants used in sintering experiments have historically excluded perchlorate because it is a fire hazard in the laboratory setting. When researchers have added it deliberately ,at concentrations matching measured Martian surface values of around half a percent by weight ,they find that perchlorate decomposes between roughly 400 and 600 degrees Celsius, well below sintering temperature, releasing oxygen gas inside the compacted material before the surrounding silicate matrix has consolidated enough to contain or vent it cleanly. The result is internal porosity, potential cracking, and structural weakness in the sintered product. Managing this gas evolution is an unsolved process variable in Martian regolith sintering that the construction materials community is actively working on, and it has no clean solution when starting from bulk unprocessed regolith at Martian perchlorate concentrations.

Electrostatic pre-treatment of the feedstock addresses this problem as a direct consequence of the oxygen production process already occurring. The silicate fraction delivered by the separator to the construction pathway has had its perchlorate content reduced by whatever enrichment factor the separation achieves. The degree of perchlorate reduction depends on separation efficiency, which is a measurable parameter to be established during precursor missions. Material sintered from this pre-treated fraction faces a substantially reduced gas evolution problem during firing, with correspondingly better control over porosity and structural integrity in the final product. The two functions ,perchlorate concentration for oxygen production, and perchlorate removal for construction quality ,are served by the same separation step. The integration requires no dedicated additional processing hardware and minimal incremental power beyond that already allocated to the separation stage. The architecture produces cleaner construction feedstock automatically, as a consequence of operating for its primary resource production purpose.

There is a further connection worth noting. The calcium chloride residue from the perchlorate thermal decomposition reactor ,already discussed as a hygroscopic water recovery agent ,is also a flux material in silicate systems. Calcium chloride in contact with silicate minerals at elevated temperatures lowers the effective softening point of the mineral mixture through eutectic behavior, meaning the pre-treated silicate fraction combined with a controlled addition of calcium chloride residue could sinter at temperatures somewhat below those required for pure silicate material. The energy saving from this lower sintering temperature, while modest per batch, compounds over the operational lifetime of the facility and reduces overall power demand for construction. Whether this effect is large enough to matter in the specific mineralogy of a given landing site is a question that Phase 2 sample characterization and small-scale sintering tests would resolve before any production commitment is made.

The reactor stage

Perchlorate thermal decomposition is straightforward chemistry. Calcium perchlorate decomposes above approximately four hundred and twenty degrees Celsius, releasing molecular oxygen and leaving calcium chloride as the solid residue. The thermodynamics are well established, the stoichiometry is fixed, and the engineering challenges are thermal management and gas capture rather than chemical uncertainty. The energy requirement for the decomposition reaction itself is on the order of four hundred kilojoules per mole of calcium perchlorate, which for a feed stream at one or two percent perchlorate by weight translates to thermal energy demands that are modest relative to the solid oxide electrolysis system operating in parallel. A rough estimate places the thermal energy requirement per kilogram of oxygen produced through this pathway at two to three times higher than atmospheric electrolysis, primarily because the feed stream must be heated to decomposition temperature whether or not the perchlorate concentration in the feed is high, and the silicate matrix absorbs heat without contributing to product.

This energy penalty makes perchlorate decomposition less efficient per unit of oxygen produced than atmospheric electrolysis under most conditions. It does not make the pathway uninteresting. The atmospheric electrolysis system depends entirely on one feedstock ,the atmosphere ,and operates best as a continuous process drawing steady power. The perchlorate reactor can be operated in batches, scheduled during periods when primary systems are at lower power demand, using a different feedstock that is available regardless of atmospheric conditions. The two pathways provide operational redundancy and different failure modes. They also share oxygen storage and compression infrastructure, which is an integration benefit.

The calcium chloride residue from the reactor is not wasted. It is a hygroscopic salt that will absorb atmospheric water vapor when exposed to the Martian atmosphere during the nighttime hours when temperature drops and relative humidity increases. This water absorption requires no active processing and no additional feedstock ,it occurs passively as a consequence of the reactor's operation. The absorbed water is recovered by heating the residue to approximately one hundred and fifty degrees Celsius, which liberates adsorbed and deliquesced water as vapor while leaving structural mineral water, which requires much higher temperatures to release, bound in the solid phase. The recovered vapor condenses and is collected.

The water yield from this recovery process is not large. At Martian nighttime humidity conditions in mid-latitude regions, laboratory analog work suggests absorption rates on the order of five to ten grams of water per kilogram of calcium chloride per diurnal cycle under favorable conditions. At a reactor throughput producing a few kilograms of residue per day, the daily water recovery might be measured in tens of grams. This is a supplementary stream, not a primary water supply. Its value is that it produces water as a byproduct of operations already occurring, at low additional hardware mass and near-zero additional power cost.

Power scheduling and integration with atmospheric processing

The complete electrochemical stack interfaces with the primary atmospheric electrolysis system at two points: power and oxygen storage. The power interface is the more constraining.

NASA's Fission Surface Power program, which completed Phase 1 concept designs for a forty-kilowatt electrical system and has issued solicitations for Phase 2 development targeting a lunar demonstration in the early 2030s, represents the current baseline for surface power on human Mars missions. More recent program direction has updated the target to at least one hundred kilowatts for the operational system, with the forty-kilowatt class representing the near-term demonstration milestone. A production-scale atmospheric electrolysis system at two kilograms of oxygen per hour requires approximately twenty-five to thirty kilowatts of continuous power. The electrochemical stack described here adds an estimated five percent to that demand in continuous operation, with higher instantaneous draws during batch thermal processing that can be scheduled during periods of lower primary system load.

The practical scheduling picture is that atmospheric electrolysis runs best continuously, benefiting from steady thermal management and avoiding the startup energy penalties of cycling the solid oxide cells through large temperature swings. The perchlorate reactor runs best in batches, is tolerant of scheduling around peak power availability, and can take advantage of waste heat from the electrolysis system if thermal plumbing connects the two units. The collector array runs passively during dust season and draws power only for electrostatic separation of accumulated material, which can be scheduled whenever power margin exists. The water recovery heating occurs during daylight when thermal energy is most available and power demand on life support systems is lower.

None of these scheduling constraints require hardware that does not already exist in concept. They do require a power management and scheduling system that treats the integrated stack as a single unit with time-varying load profile rather than a collection of independent consumers. That is an integration engineering problem, not a physics problem.

Chapter 28: The Demonstration Ladder

The architecture proposed in the preceding chapter spans a wide range of technology readiness levels. Some components have operated on Mars. Others exist as laboratory demonstrations. Others are conceptual integrations of separately proven physics. Treating all of these as equally ready for flight would be an engineering error. What is needed is a structured sequence that advances the lower-readiness components while building on the foundation of what is already proven, in an order that reflects both the technical dependencies and the mission opportunities available in the next decade.

The following sequence is not a mission plan. It is a de-risking ladder: a description of what needs to happen, in what order, before the integrated architecture can be trusted to support a human crew. Each phase builds on the phase before it. Skipping phases is possible in principle but transfers risk forward rather than eliminating it.

Phase zero: orbital characterization

Before any surface hardware is committed to a specific site, the orbital dataset needs to be used more deliberately than it has been. The Context Camera archive from Mars Reconnaissance Orbiter contains years of dust devil detections whose geographic and seasonal structure has been partially analyzed but not systematically applied to site selection. The SHARAD radar sounder has mapped subsurface ice distribution across a substantial fraction of the planet's mid-latitudes but the mapping has not been systematically correlated with potential landing sites for human missions. The CRISM spectral database contains mineralogical information that could constrain surface oxidation state and perchlorate distribution patterns but has not been mined for those specific parameters at candidate sites.

Phase zero is not a new mission. It is an analysis campaign using existing data. The output is a ranked set of candidate sites characterized not only by ancient geology but by dust activity index, estimated perchlorate surface concentration based on spectral proxies and regional activity history, subsurface ice depth and estimated thickness, and landing hazard assessment. This analysis can be completed before the next Mars surface mission launches and would provide the selection basis for Phase 1 hardware placement that does not currently exist.

Phase one: targeted robotic demonstration

The first surface mission in the ladder has a specific and limited objective: demonstrate whether electrostatic collection of ambient Martian dust produces meaningful perchlorate enrichment relative to bulk surface regolith. This is the single most important unknown in the proposed architecture. Everything downstream of the collector depends on it. If enrichment is negligible, the collector adds mass for no resource benefit, and the architecture must be redesigned around conventional excavation. If enrichment is significant, the case for passive collection as a feedstock strategy becomes substantially more credible.

A Phase 1 payload could be modest. A compact collector array of a few square meters, a chemical analyzer capable of measuring perchlorate concentration in collected fines with sufficient precision to detect differences from bulk surface measurements, and a meteorological package to correlate collection rates with local vortex activity would together occupy less than twenty kilograms and could be accommodated on a small lander or as a secondary payload on a larger mission. The mission should be placed at a site that Phase 0 analysis has characterized as moderately active ,not in the highest activity zone, where results would represent an optimistic outlier, and not in the lowest activity zone, where the feedstock rate would be too low to measure anything ,but at a site representative of conditions at plausible human mission landing sites.

The result of Phase 1 is binary in the most useful sense. Either electrostatic collection enriches perchlorate, in which case the Phase 2 thermal processing demonstration is justified. Or it does not, in which case the architecture is revised to treat dust as a hazard to manage rather than a resource to harvest, and the oxygen production strategy relies entirely on atmospheric electrolysis and perchlorate excavated through conventional means. A negative result from Phase 1 is not a failure of the overall resource strategy. It is a clarification of which pathway to invest in, and it is better to have that clarification from a ten-kilogram demonstration payload than from a failed production system on a crewed mission.

Phase two: integrated robotic resource mission

Assuming Phase 1 demonstrates meaningful enrichment, Phase 2 is the first test of the integrated stack as a system. A dedicated resource demonstration lander carries the full sequence: collector array, two-stage electrostatic separator, thermal decomposition reactor, and hygroscopic water recovery tray. It also carries meter-scale drilling capability, both to test the subsurface access hardware that human missions require and to collect depth-profiled samples that directly test the oxidation gradient hypothesis developed in the first half of this book.

Phase 2 operates autonomously for at least one full Mars year before any crew interaction. The output is operational performance data for each stage of the stack under actual Mars conditions, including the failure modes and degradation patterns that laboratory testing cannot predict. It is also an oxygen and water inventory ,small quantities, but quantities produced by the actual system that will later be scaled for human use, under the actual conditions where that scaled system will operate.

Phase 2 is where the performance numbers either close or do not. The mass budget assumptions, the power scheduling model, the thermal management of the reactor, the water uptake rates in the hygroscopic recovery stage ,all of these are estimates based on physics and analog testing until a real system runs on Mars and produces measured results. Phase 2 converts estimates into operational data. That conversion is the engineering prerequisite for crew dependence on any component of the stack.

Phase three: crew support integration

Phase 3 is the first crewed mission to incorporate the electrochemical stack as part of its life support and propellant production architecture. By this point, the stack has operated autonomously on Mars for at least one Mars year, its performance has been characterized, and its failure modes have been addressed in design. The primary oxygen production system ,atmospheric solid oxide electrolysis ,has been running at full scale since before crew arrival, using the forty-to-one-hundred-kilowatt fission surface power system as its energy source. The electrochemical stack supplements that system rather than replacing it.

The crew arrives at a site where oxygen production is already running, where some water has been extracted and stored, and where the collection and processing hardware has been operating long enough to have accumulated significant feedstock. The first human use of these systems is verification and maintenance, not primary dependence. Primary dependence comes in later missions, after the systems have demonstrated the reliability required for life safety applications.

This phasing is not conservatism for its own sake. It reflects the engineering reality that no system can be trusted for life safety applications without operational heritage, and that operational heritage on Mars can only be accumulated by operating on Mars. The fastest route to a crewed mission that can genuinely rely on in-situ resources is not to design those systems in the abstract but to fly the precursors early enough that the data they generate can inform the crewed mission design.

Chapter 29: The Nitrogen Gap

The engineering section of this book has described systems for producing oxygen, extracting water, generating power, and manufacturing construction materials from Martian resources. A reader working through the architecture might reasonably conclude that the primary life-support requirements for a human Mars surface mission are addressed ,oxygen from atmospheric CO_2 electrolysis, water from subsurface ice or hygroscopic extraction, power from fission surface reactors supplemented by solar. What that reader would be missing is the second gas that humans need to breathe, and the one for which no clean solution currently exists at Mars surface conditions.

Nitrogen is approximately 78 percent of Earth's atmosphere by volume, and its role in a breathable atmosphere is not incidental. Pure oxygen at partial pressures sufficient for human respiration ,roughly 20 to 23 kilopascals, or about 3 psi, equivalent to breathing at roughly 3,000 metres altitude ,cannot be sustained indefinitely in a habitat without a diluent gas, for two distinct reasons. The first is physiological. Breathing elevated-concentration oxygen over extended periods causes oxygen toxicity, damaging lung tissue and the central nervous system at sufficient exposure levels. The second is operational. Habitats with elevated oxygen concentrations are substantially more susceptible to fire ,the ignition threshold for materials drops, flame propagation speeds increase, and fires that would self-extinguish in normal air can spread lethally in an oxygen-enriched environment. Both of these constraints ruled out pure oxygen habitats for long-duration spaceflight after the Apollo 1 fire in January 1967, which killed three astronauts in a 100 percent oxygen cabin at slightly above ambient pressure on the launch pad. The ISS operates at essentially sea-level atmospheric composition, near 101 kilopascals total pressure with oxygen at approximately 21 percent. Robert Zubrin, in The Case for Mars, has argued for a lower-pressure Mars habitat atmosphere of roughly 34 kilopascals total ,approximately 24 kilopascals oxygen and 10 kilopascals nitrogen ,which reduces structural requirements for the pressure vessel while maintaining physiological adequacy. The specific numbers are debatable, but the requirement for a substantial nitrogen partial pressure as a diluent is not. A habitat running on oxygen alone is an unacceptable fire risk over a multi-year surface stay.

The Martian atmosphere contains nitrogen at approximately 2.6 percent by volume. At surface pressure of roughly 600 to 700 pascals, that translates to a nitrogen partial pressure of around 16 to 18 pascals ,less than 0.02 kilopascals. A habitat requiring 10 kilopascals of nitrogen partial pressure needs to concentrate atmospheric nitrogen by a factor of roughly 500 to 600 relative to the ambient Martian atmosphere. This is not achieved by simply pumping Martian air into the habitat; the CO_2 that makes up 95 percent of the atmosphere must first be removed, and the remaining nitrogen and argon mixture must be concentrated to useful pressures. Cryogenic separation ,liquefying the CO_2 by cooling, leaving nitrogen and argon behind as gases ,works in principle and has been studied in the context of Mars ISRU since at least the 1990s. NASA technical work on continuous processing of Martian atmospheric gases noted that the nitrogen and argon present in the atmosphere actively interfere with CO_2 capture processes, blocking adsorption beds and slowing freezing approaches, meaning that atmospheric processing for MOXIE-type oxygen production already must deal with nitrogen as a nuisance ,separating it out rather than concentrating it. If that separated nitrogen stream were captured and stored rather than vented, it would become the nitrogen reserve a habitat needs. The engineering for this exists in concept but has not been demonstrated in an integrated system, and it represents a mass and power cost that does not appear in the architecture described in Chapter 27.

The fertiliser dimension of the nitrogen problem compounds the supply challenge. Plant growth requires fixed nitrogen ,nitrogen atoms bonded to other elements, typically as ammonia (NH_3), nitrate (NO_3^-), or urea ,rather than molecular nitrogen gas (N_2). Earth agriculture relies on the Haber-Bosch process, which fixes atmospheric nitrogen with hydrogen over an iron catalyst at high temperature and pressure, and on biological nitrogen fixation by root-associated bacteria in leguminous plants. Neither approach transfers straightforwardly to early Mars operations. Haber-Bosch at small scale is energy-intensive and requires hydrogen feedstock that is itself scarce on Mars until water extraction is well established. Biological nitrogen fixation requires plant root systems, specific bacterial strains, and soil chemistry that would take considerable time to establish in Martian regolith even after perchlorate remediation. Yet without fixed nitrogen, food production from local agriculture is not possible. Every crop plant requires nitrogen in its biomass ,typically 15 to 50 grams of nitrogen per kilogram of dry plant mass ,and a crew that cannot grow food locally remains indefinitely dependent on Earth supply for calories.

This is where a development from European plasma physics research becomes directly relevant to the engineering framework of this book. In 2022, independent groups at the University of Antwerp and the University of Lisbon demonstrated that non-thermal plasma reactors operating in simulated Martian atmospheric composition ,96 percent CO_2, 2 percent N_2, 2 percent argon ,could simultaneously produce oxygen through CO_2 dissociation and fix nitrogen through oxidation of the N_2 fraction to nitrogen oxides. The University of Antwerp group, using microwave plasma excitation at an energy input of 1 kilowatt-hour ,equivalent to MOXIE's operational energy draw ,achieved oxygen production at 47 grams per hour and nitrogen oxide production at 1.25 grams per hour, representing approximately 7 percent fixation of the available N_2 fraction in the feed gas. At those production rates, one kilowatt of continuous power devoted to plasma atmospheric processing would yield roughly 1.1 kilograms of oxygen per day alongside approximately 30 grams of nitrogen oxides per day.

The nitrogen oxides dissolve in water to form nitric acid, which can be neutralised with a base ,ammonia, if available, or calcium carbonate from Martian regolith ,to produce nitrate salts directly usable as plant fertiliser.

The connection to this book's existing framework is not incidental. Non-thermal plasma operating on Martian atmospheric gas is the same category of physics as triboelectric discharge in dust devils ,in both cases, electrical energy excites CO_2 and its associated trace gases into reactive plasma states where molecular bonds break and reform into new chemical species. The Martian atmosphere at surface pressure is, as Vasco Guerra of the University of Lisbon noted explicitly in the published work, operating at essentially the ideal pressure window for non-thermal plasma chemistry. Electron mean free paths are long enough for efficient energy coupling, collision frequencies are high enough for useful reaction rates, and the low ambient temperature ,averaging roughly 210 Kelvin on the surface ,slows the back-reaction that recombines products into starting materials, increasing net conversion efficiency compared to what the same reactor achieves at Earth temperatures. Mars does not just tolerate plasma chemistry. Its surface conditions favour it. The dust devil electrochemistry this book describes is one natural expression of that favourability. Engineered plasma reactors are the deliberate exploitation of the same physical environment.

The plasma nitrogen fixation approach has genuine advantages over alternatives for early missions. Cryogenic nitrogen separation from the Martian atmosphere is a competing pathway that concentrates the 2.6 percent nitrogen fraction by removing CO_2. Plasma fixation operates differently ,rather than concentrating the trace nitrogen, it converts it in place during the oxygen production process, producing a useful chemical product from what would otherwise be a process contaminant. The two approaches address different needs: cryogenic separation provides molecular nitrogen gas for habitat pressurisation, while plasma fixation provides reactive nitrogen compounds for fertiliser. A complete nitrogen strategy for a Mars surface base probably requires elements of both, with the balance depending on whether food production or habitat atmosphere is the more pressing near-term requirement and on the relative mass and power costs of each approach as hardware matures.

The current technology readiness level of plasma nitrogen fixation for Mars is low. The laboratory demonstrations at Antwerp and Lisbon were conducted in bench-scale reactors under simulated Martian atmospheric composition, not in hardware designed or qualified for space flight. The 2025 ACS Sustainable Chemistry and Engineering review of plasma ISRU for Mars characterises the technology as promising but noting real deployment challenges including thermal stress resistance of reactor components, dust ingestion ,since any atmospheric intake on Mars carries dust that will contaminate plasma chambers ,and the requirement for efficient gas separation downstream to isolate the oxygen and nitrogen oxide streams from each other and from unreacted feed gas. Integrating this technology into the electrochemical architecture described in Chapter 27 would require an atmospheric intake and pre-filtration stage, a plasma reactor module operating in parallel with or as an alternative to the solid oxide electrolysis system, and a downstream water scrubber to convert nitrogen oxides to nitrate solution. Each of these components has terrestrial precedent, none has been flight-qualified for Mars, and the integrated system does not yet exist in any form. A realistic TRL assessment places it at 3 ,the basic concept and underlying physics are demonstrated in the laboratory, but no relevant environment testing of an integrated system has been performed.

What should go into the Phase 1 and Phase 2 demonstration ladder described in Chapter 28 is at minimum a measurement of the nitrogen oxide yield achievable from plasma processing of actual Martian atmosphere under actual Mars surface conditions, alongside the oxygen production measurements already planned. This requires adding a nitrogen oxide detector to the Phase 1 payload ,a modest addition in mass, since electrochemical nitrogen oxide sensors are small and well-developed for industrial air quality monitoring ,and a water absorption column to test whether the yield is sufficient to produce measurable nitrate concentrations in a few hundred millilitres of water over a demonstration period of weeks to months. The result would establish whether the laboratory production rates translate to the actual Martian environment, and whether the architecture can be extended to provide a nitrogen stream alongside its oxygen and water outputs. A negative result ,negligible nitrogen oxide production under real Mars conditions ,would indicate that the cryogenic separation pathway must carry the full nitrogen burden and should be sized accordingly in mission planning. A positive result would identify a pathway to producing crop fertiliser from the same atmospheric processing infrastructure already justified for oxygen production, with essentially no additional hardware mass beyond the detection and scrubbing components.

The nitrogen gap is real and it affects every human Mars mission at a level of consequence equal to water and oxygen. It has been noted in the Mars engineering literature for decades, typically in a paragraph that acknowledges the problem and defers to future solution. The plasma fixation pathway is the first candidate that connects directly to hardware already being developed for oxygen production and to the atmospheric plasma physics that this book's scientific framework describes as a fundamental feature of the Martian environment. That connection does not resolve the engineering challenge ,the hardware does not yet exist, the integrated system has not been demonstrated, and the yield numbers from small laboratory reactors have not been validated in flight. But it identifies where the solution is most likely to come from, and it establishes that nitrogen and oxygen production may, under the right plasma reactor design, be products of the same atmospheric processing operation rather than independent systems competing for the same limited mass and power budget.

Chapter 30: Failure Modes and What They Would Mean

An engineering proposal that does not address its own failure modes is not an engineering proposal. The electrochemical architecture described in this section could fail at several points, and identifying those points in advance is more useful than discovering them during a crewed mission. More than that: knowing in advance what would falsify the architecture allows each phase of the demonstration ladder to be designed with the right measurements to catch failures early, when they can still be addressed.

Electrostatic enrichment may not be significant

The most consequential potential failure is that electrostatic collection of ambient Martian dust does not produce meaningful perchlorate enrichment relative to bulk surface material. This would occur if the triboelectric charging of perchlorate-bearing particles is not sufficiently different from that of silicate particles to produce useful separation under Mars surface conditions, or if atmospheric mixing homogenizes the dust before it reaches a collection surface, or if the dominant fine fraction of Martian dust simply does not carry the perchlorate concentration that bulk regolith measurements suggest.

If Phase 1 measurements show that collected dust has perchlorate concentrations within measurement uncertainty of uncollected surface material, the passive collection pathway should be abandoned. The architecture then relies entirely on atmospheric electrolysis for oxygen production and on conventional excavation or ice extraction for water. Both of those pathways are viable independently. The electrochemical stack would be reduced to its atmospheric electrolysis component plus whatever mechanical extraction capability is available. This is not a mission failure. It is a design simplification.

Dust throughput may be too low at realistic sites

Even if electrostatic enrichment is significant, the feedstock rate at sites suitable for human missions may be insufficient to justify the mass of collection and processing hardware. Dust activity varies by more than an order of magnitude across the Martian surface, and human mission landing sites are constrained by elevation, landing hazard, and resource access requirements that do not select for maximum dust activity. A site at which the expected annual collection rate is ten kilograms of enriched dust rather than a hundred kilograms produces proportionally less oxygen and water from the electrochemical pathway, and at some threshold, the mass of the collection hardware exceeds the mass of the resources it would recover.

Calculating that threshold requires Phase 0 dust activity characterization at candidate sites and Phase 1 collection rate measurements at a representative site. If the numbers do not close at the planned landing site, the collection array is either reduced in size or eliminated from the mission manifest. The atmospheric electrolysis system is unaffected. The kill criterion here is a positive calculation ,the system makes sense to fly when and only when the resource return exceeds the mass and power cost of the hardware over the mission duration.

Thermal decomposition fouling

Calcium perchlorate decomposes cleanly in laboratory experiments at controlled feed composition. Real Martian regolith contains sulfates, perchlorates of multiple cation species, iron oxides, and silicates in proportions that vary by site and depth. Multi-component feed materials can produce decomposition products that coat reactor walls, clog gas outlets, or alter the thermal profile of the reactor in ways that reduce efficiency or require active intervention over mission timescales. This class of failure is well documented in terrestrial chemical processing and is the reason that process industries run extended pilot campaigns before committing to production scale.

The mitigation is straightforward in concept: design the reactor for cleaning access and for modular replacement of the components most likely to accumulate deposits, and characterize the actual regolith composition at the mission site before committing to a reactor design calibrated for that feed. Phase 2 operation provides the extended runtime data needed to characterize fouling behavior before crew dependence on the system. If fouling rates are acceptable over one Mars year of operation, the Phase 3 crewed mission can proceed. If fouling rates require more frequent maintenance than the crew schedule can support, the reactor design is revised before the crewed mission.

Hygroscopic water recovery below economic threshold

The water recovery subsystem depends on calcium chloride and related hygroscopic residues absorbing meaningful quantities of atmospheric water vapor at Mars nighttime humidity levels. If actual uptake rates at the mission site are at the low end of laboratory estimates ,perhaps two or three grams per kilogram per diurnal cycle rather than five to ten ,the system produces water at rates that may not justify the thermal energy cost of recovery. The hygroscopic recovery subsystem is the lowest-cost component of the stack in mass terms, but even a small mass commitment requires justification if the output is negligible.

This failure mode is the most likely of those discussed here, because Mars atmospheric humidity is highly variable and because laboratory analog testing of water uptake may not fully capture the dynamic conditions of a real Martian diurnal cycle. The consequence of confirming low uptake rates in Phase 2 is simply that the recovery tray is not included in the Phase 3 mission manifest. Primary water supply shifts entirely to the ice extraction system. The rest of the architecture is unaffected.

Electrostatic hazards near habitats and avionics

The confirmation that Martian dust events produce actual electrical discharges creates a hazard category that was previously inferred but not demonstrated. Fifty-five recorded discharge events across Perseverance's mission record establishes that arcing occurs, but does not establish the rate at which arcs of sufficient energy to damage electronics or ignite oxygen-enriched atmospheres can be expected near a human habitat over a multi-year mission. The collector array, which deliberately concentrates charged material and maintains high-voltage electrodes, amplifies this risk locally.

Engineering responses include physical separation of the collector array from habitats and oxygen processing equipment ,the array should be placed upwind and at a safe distance, with separation distances derived from the electrostatic discharge energy characterization that Perseverance has begun and future missions should continue. Grounding all conductive structures connected to the habitat to a common reference prevents charge accumulation on structures adjacent to the collector. The collector's high-voltage power supply should incorporate arc detection and automatic shutdown, a standard precaution in terrestrial electrostatic processing equipment. Oxygen processing lines must not pass through the high-field region of the collector array.

None of these precautions are novel. They are standard practice in terrestrial electrostatic processing facilities. Their application to Mars requires characterization of the ambient discharge environment at the specific mission site, which is a measurement that can be made by the Phase 1 demonstration payload at minimal additional cost.

Chapter 31: Site Selection as an Engineering Decision

The scientific case for incorporating modern atmospheric chemistry into Mars landing site selection has been developed across the preceding chapters. The engineering case is different in emphasis, because it must satisfy constraints that do not appear in the scientific literature. A landing site that maximizes biosignature preservation potential or perchlorate resource concentration is not useful if it cannot be reached safely, powered adequately through the worst expected dust storm, or equipped with the subsurface access needed for the mission's resource goals. Site selection for early human missions is an optimization problem with multiple coupled variables, and atmospheric chemistry is one of them.

The constraints that eliminate most of Mars

Entry, descent, and landing systems impose an elevation floor on accessible sites. Atmospheric deceleration depends on the column of air above the landing site; lower elevation means denser atmosphere and more effective aerobraking. The practical consequence is that most sites above roughly negative one kilometer in MOLA elevation are difficult to reach with current EDL system designs at human payload masses. This immediately excludes the highest-interest ancient geological terrains on the Tharsis plateau and much of the southern highlands.

Landing hazard assessment eliminates sites with high boulder density, steep slopes, or terrain roughness at scales that would challenge the landing system's adaptive capability. Ancient lake beds and delta deposits, which rank highest for biosignature science, tend to occur in topographically complex terrain near channel systems and rim structures. The scientifically optimal sub-regions within these deposits are often not the operationally safe landing zones. Detailed landing hazard mapping from high-resolution imaging is a prerequisite for committing to any specific site.

Solar power viability at sites not initially equipped with fission surface power requires insolation characterization through the worst expected dust storm season. Regional dust storm events can reduce solar insolation by factors of two to five for periods of weeks to months. A site that appears viable under average conditions may be inoperable under storm conditions. Until fission surface power is flight-proven on Mars, any site selected for a human mission must either demonstrate acceptable power margin through documented storm events or carry sufficient energy storage to bridge storm periods at minimum life-critical load.

What the dust activity dataset adds

Within the sites that survive the basic engineering filters, dust activity characterization provides information relevant to at least three mission design decisions that are currently made without it.

The first is the solar panel degradation rate. A site in a high dust-devil-activity zone accumulates panel dust at higher rates than a quiescent site, reducing effective power generation and increasing the maintenance burden of panel cleaning. Quantitative dust accumulation rate estimates at candidate sites, derived from dust devil track density and vortex size distribution in the Context Camera archive, would allow solar panel degradation to be factored into power system sizing rather than treated as a background uncertainty.

The second is electrostatic collection potential. If the Phase 1 demonstration confirms that passive collection enriches perchlorate, then a site's dust activity level directly determines the throughput of the collection subsystem. A site in the seventy-fifth percentile of dust activity for accessible landing locations may collect two or three times the annual feedstock of a site in the twenty-fifth percentile. At the mass efficiency levels that early human missions require, that difference is significant.

The third is electrostatic discharge hazard characterization. Sites with high dust activity will have more frequent and potentially higher-energy discharge events than quiescent sites. The safety implications for habitat grounding design, electronics protection, and oxygen processing proximity are site-specific and should be quantified before hardware is committed to a particular location.

The preservation variable and drill depth requirement

For missions with astrobiology objectives, the site selection framework developed in this book's earlier chapters has a direct engineering translation. Any mission that plans to search for intact organic biosignatures should specify the depth at which modern oxidative alteration becomes negligible at the specific site it selects, and should then design its sample acquisition hardware to reach that depth. Specifying the sample acquisition hardware to reach seven centimeters because that is what existing drill designs achieve, and then hoping that unaltered material happens to be present in the top seven centimeters, is not a scientific strategy. It is a hardware limitation being treated as though it were an acceptable constraint.

The JPL Mars Deep Drill system targets one hundred meters of penetration in a payload mass under one hundred kilograms. At a far smaller scale, a one-to-three-meter coring capability would be sufficient to characterize the oxidation gradient at most sites and access material substantially below the zone of rapid modern alteration. The Planetary Volatile Extractor concept demonstrates meter-scale access in ice-bearing regolith. These capabilities exist as hardware at varying stages of readiness. The engineering decision is whether to include one of them in the mission manifest or to accept the preservation limitation that comes from sampling only the near surface.

The argument for including meter-scale access is not primarily scientific. It is economic. If a mission is going to Mars to search for biosignatures, the probability that it finds them is a function of whether it is sampling material that preserves them. A drill that reaches material with a high probability of preservation costs more than a drill that does not, but it delivers better return on the entire mission investment. Site selection and sampling hardware design are not independent decisions. They should be made together, with explicit acknowledgment of the depth at which the specific site's alteration profile makes surface sampling inadequate.

Toward a formal scoring approach

Site selection for human Mars missions is currently conducted through expert evaluation of multiple parameters without a formal weighting scheme. Different mission planning teams weigh ancient geology, landing safety, resource access, and operational convenience differently, and the process does not always make those weights explicit. This creates inconsistency across mission planning cycles and makes it difficult to compare candidate sites on equal terms.

A formal scoring approach need not be rigid. It would include, at minimum, elevation above the EDL floor, landing hazard index derived from high-resolution imaging analysis, insolation margin through the documented worst dust storm season at that latitude, proximity to mapped subsurface ice within drilling range of the planned capability, dust activity index from the Context Camera archive, and estimated perchlorate concentration based on spectral proxies and regional activity correlation. Each parameter would carry an explicit weight reflecting the mission's priorities, and the weights would be disclosed as part of the site selection rationale so that they can be reviewed and revised as mission objectives evolve.

Adding dust activity index to this parameter set does not complicate the evaluation. It uses data that already exists. It captures information that is currently handled informally, if at all. And it connects the scientific framework developed in the first half of this book to the engineering decisions that determine where that science actually gets done.

Chapter 32: The Mars Gradient Sampling System

The demonstration ladder described in Chapter 28 is a sequence for validating the electrochemical architecture ,the resource extraction stack that the first human Mars missions will depend upon. It is not a sequence for validating the scientific framework that motivated the architecture. Those are different problems. The resource architecture can be confirmed or refuted by measuring what it produces: oxygen yield, water recovery rate, perchlorate enrichment factor. The scientific framework ,the claim that Mars has a spatially structured oxidation gradient driven by atmospheric electrochemistry, and that this gradient determines what any mission will actually find when it drills ,requires a different kind of measurement entirely. It requires a measurement that no instrument flown to Mars has been designed to make.

Every subsurface measurement taken on Mars to date has been made at a single point, at a single location, with a single drill that reaches seven centimetres into consolidated rock. The measurements are not wrong. The instruments are well-designed for what they were asked to do. But they were asked to characterise a point, and the framework developed in this book is a framework about spatial structure. You cannot test a gradient with a point measurement any more than you can test a temperature distribution with a single thermometer. The measurement tool and the scientific question are mismatched, and they have been mismatched for the entire history of Mars surface exploration.

The Mars Gradient Sampling System is a proposed instrument designed specifically to close that mismatch. It is not an incremental improvement to existing Mars drilling capability. It is a purpose-built field experiment ,a multi-node, multi-scale simultaneous measurement of subsurface chemistry, atmospheric electric forcing, and structural tomography at a single Mars surface site ,that would test the oxidation gradient hypothesis directly for the first time. The distinction between an instrument and an experiment matters here. An instrument measures. An experiment tests a hypothesis. The MGSS is designed as an experiment, with a specific binary outcome: the gradient framework is confirmed or it is refuted, and either result is scientifically productive.

The measurement that has never been made

Before specifying the hardware, it is worth being precise about what measurement is actually needed. The scientific argument of this book makes three connected claims. First, that modern triboelectric atmospheric chemistry produces oxidants continuously at the Martian surface and that the accumulation of those oxidants decreases with depth below the surface in a gradient whose shape depends on local atmospheric activity history, rock permeability, and exposure duration. Second, that this gradient varies laterally across the surface in ways that correlate with dust devil activity ,high-activity sites accumulate oxidants faster than low-activity sites, and the gradient is steeper and shallower at high-activity sites. Third, that the existing perchlorate measurements at Phoenix and Curiosity represent a dynamic steady state between triboelectric production and radiolytic destruction, and that this steady state concentration is being actively maintained at the present day rather than inherited from ancient processes.

Each of these claims is supported by published laboratory work, atmospheric physics, and orbital observations. None of them has been tested by a direct field measurement on the surface of Mars. The gap between the supporting evidence and the claim is not large ,the physics is well-established and the indirect evidence is consistent ,but it is a gap, and in planetary science a gap of that kind is where competing explanations live. Testing the framework requires measuring three things simultaneously at the same location: the subsurface chemistry profile from the surface to at least one metre depth, the spatial variation of that profile across lateral distances from one metre to ten metres, and the local atmospheric electric forcing that the framework predicts drives the chemistry. Without all three, the measurement is incomplete. A vertical profile without lateral comparison cannot distinguish local geological heterogeneity from the atmospheric chemistry gradient. A lateral comparison without atmospheric forcing data cannot establish causation. Atmospheric forcing data without the chemistry is atmospheric without consequence.

The MGSS measures all three simultaneously. That simultaneity is not a design luxury ,it is a scientific requirement.

Architecture: the Pathfinder configuration

The MGSS-Pathfinder is the first-flight configuration of the system, designed for a Discovery-class lander or as a primary secondary payload on a larger mission. It deploys four nodes on the Martian surface: a central node designated C0 and three outer nodes designated N1, N2, and N3, positioned on three radial arms separated by 120 degrees. The outer nodes are positioned at five to ten metres from the centre, a baseline chosen to capture the spatial scale at which dust devil tracks leave measurable surface chemistry signatures ,the scale at which the lateral gradient is predicted to be most readily detectable in a single Mars year of accumulation.

The central node carries the meteorological mast, the ground-penetrating radar transmitter, the shared spectroscopic sonde base station and winch, and its own drill unit and electric field sensor. One of the three outer nodes ,designated N2 ,carries the sealed atmospheric exclusion chamber in addition to its drill unit and electric field sensor. The remaining two outer nodes ,N1 and N3 ,are standard open drill nodes with electric field sensors and ground-penetrating radar receivers. Surface cable runs from C0 to each outer node carrying power, data, and the sonde deployment line.

This arrangement of four nodes is not a simplification of a more complex ideal system. It is the minimum configuration capable of answering the primary scientific question. Three outer nodes are needed rather than one or two because the azimuthal variation in chemistry between the arms ,which may differ depending on the prevailing wind direction and its relationship to the local dust devil source regions ,is a measurement in itself, and distinguishing azimuthal variation from radial variation requires observations on multiple arms. The sealed node must be one of the outer nodes rather than the central node because the comparison between sealed and open chemistry is most informative when the two are separated laterally ,demonstrating that the sealed node's chemistry diverges from its geologically identical neighbour over time, not merely that it differs from a remote location.

The drill units

Each node is equipped with a rotary-percussive drill unit adapted from the TRIDENT system developed by Honeybee Robotics, which reached technology readiness level six through development for the VIPER and PRIME-1 lunar missions. The TRIDENT configuration relevant to this application uses a 25.4 millimetre diameter tungsten carbide-tipped auger on a one metre linear deployment rail, with separate rotary and percussive actuators each rated at 200 watts peak power draw. Rotation speed reaches 120 revolutions per minute, and the percussive system delivers two joules per blow at rates up to 972 blows per minute, actuated on demand rather than continuously. The system demonstrated the ability to drill eight holes to nearly one metre depth in six days of field testing at Haughton Crater under autonomous control, successfully recovering from five recorded fault states without external intervention.

The MGSS adaptation of this heritage differs from the TRIDENT flight configuration in four specific respects. First, the rover arm interface is removed and replaced with a fixed footpad anchoring frame carrying four hardened ground-penetrating spikes, sized to prevent displacement under the percussive reaction loads of drilling in a system with no counterweight beyond its own mass. The footpad is mechanically driven into the surface regolith on deployment. Second, a downhole sonde receiver port is integrated into the drill collar, allowing the wireline sonde to be deployed into the borehole independently of the drill ,the drill advances the hole, withdraws to the surface, and the sonde is then lowered for in-situ measurements at depth. Third, the wire harness is uprated for operation in a CO_2 atmosphere at Mars surface pressure, where the Paschen breakdown threshold for insulation gaps is substantially lower than in the nitrogen atmosphere for which TRIDENT was qualified on the VIPER mission. Paschen's law predicts breakdown in CO_2 at Mars pressure of around 600 to 700 pascals at electric field strengths that would not cause breakdown in terrestrial nitrogen ,any exposed insulation gap in the drill wiring that exceeds approximately three millimetres is a breakdown risk in this environment, and the harness design must be reviewed against this threshold specifically. Fourth, the autonomous fault recovery logic is extended to include CO_2 atmosphere-specific fault modes ,ice cementation at depth and perchlorate-bearing regolith that can become adhesive when warm and abrasive when cold ,that are not present in the lunar ice scenario for which the TRIDENT fault recovery software was developed.

At each node, the drill advances in bite increments of two to ten centimetres depending on regolith consolidation, bringing cuttings to the surface via the auger flutes for analysis by the sonde at the surface. After reaching full depth ,nominally one metre, achieved over multiple drilling sessions across several Martian days ,the drill withdraws completely and the sonde is deployed into the borehole for downhole measurement at depth intervals of twenty centimetres. The science return from the drill therefore includes both the surface cuttings pile, which represents a depth-averaged sample of all material brought up from the borehole, and the discrete in-situ downhole measurements at each twenty centimetre interval, which preserve the depth stratigraphy of the chemical signal.

The shared spectroscopic sonde

A single spectroscopic sonde is shared between all four nodes, deployed sequentially by a winch system at the central node that draws the sonde along surface cables to each outer node in turn. The deployment sequence over a Martian year cycles repeatedly through all four nodes, building up time-series measurements at each depth interval as the regolith conditions change seasonally. The sonde is not deployed simultaneously at multiple nodes ,the mass and complexity cost of four separate sonde instruments is not justified at Pathfinder scale, and the sequential deployment is scientifically adequate because the chemistry gradient being measured is expected to vary on timescales of years to decades, not hours.

The sonde body is 22 millimetres in diameter, fitting within the 25.4 millimetre auger bore with 1.7 millimetres of annular clearance ,sufficient for passage in consolidated regolith, but a clearance that requires site selection to confirm that the borehole walls are stable enough not to partially collapse during the interval between drilling and sonde deployment. The sonde is approximately 280 millimetres in body length and carries four measurement sections separated by internal bulkheads.

The upper section contains the microcontroller, telemetry electronics, and wireline interface. The second section carries a miniaturised Raman laser spectrometer operating at 532 nanometres, with a side-looking window that exits through the sonde wall to allow measurement of the borehole wall mineralogy without requiring physical contact with the sample. Raman spectroscopy identifies mineral phases by their vibrational fingerprints ,calcium perchlorate, magnesium perchlorate, calcium chlorate, iron oxides, phyllosilicates, and carbonates all produce distinct Raman spectra that the instrument can separate at the concentrations expected in Martian regolith. Heritage for this component comes from the Raman Laser

Spectrometer developed for the ExoMars Rosalind Franklin rover, miniaturised for the wireline deployment geometry.

The third section carries a visible to near-infrared spectrometer operating from 400 to 2500 nanometres through a second side-looking window. The VNIR measurement complements the Raman by characterising the oxidation state of iron-bearing minerals in the borehole wall ,ferric iron phases indicative of oxidative alteration produce distinct absorption features across this wavelength range that can be quantified as a proxy for cumulative oxidative dose. This measurement provides the oxidation depth profile that is the primary physical expression of the gradient the framework predicts.

The fourth section carries an electrochemical ion sensor package including ion-selective electrodes for perchlorate, chlorate, and nitrate. This component is the highest technical risk element of the sonde. Ion-selective electrode measurements in planetary environments require contact with dissolved ions, meaning either liquid water is present in the regolith at the measurement depth, or the instrument must carry a micro-pump to expose the electrode to regolith vapour or dust drawn across a wetted membrane. Neither condition is guaranteed at the target depth in Martian regolith. The ion sensors are therefore designated as secondary science ,scientifically valuable if they function, but not mission-critical. The Raman and VNIR measurements are primary and do not depend on liquid phase chemistry.

The total sonde mass including the 280 millimetre probe body, the 24-metre wireline cable, and the winch mechanism at C0 is estimated at approximately five kilograms. This estimate carries uncertainty because no miniaturised downhole Raman/VNIR combination instrument in wireline deployment geometry has yet been built and qualified.

The electric field sensor array

Each node carries a miniaturised rotating-vane electric field mill mounted on a short post at the surface, providing a continuous record of the ambient vertical electric field component at that location. The field mill design is adapted from the Harrison and Marlton brushless miniature field meter ,a 86 gram instrument measuring 40 by 40 by 80 millimetres that achieves fair-weather field resolution of approximately one millivolt per volt per metre with a time response of 0.65 seconds, drawing less than 100 milliamps at six volts. This instrument was designed for weather balloon and robotic aircraft deployment on Earth and requires adaptation for Mars ,specifically, the motor insulation must be qualified for CO_2 atmosphere at 600 to 700 pascals, where Paschen breakdown occurs at lower field strengths than in terrestrial air, and the motor must be sized to operate at Mars surface temperatures that reach minus 120 degrees Celsius at night. These are engineering problems, not physics problems, and neither is novel ,low-temperature brushless motor qualification and CO_2 atmosphere electrical qualification are both within the state of industrial practice.

The field mills at each node log the local electric field at one-second resolution. When a dust devil passes over or near a node, the field signature rises over a period of seconds to minutes, reaches a peak correlated with the vortex passing closest to the sensor, and decays back to the ambient background. The electric field magnitude alone is not a direct measure of discharge energy;breakdown events release energy in discrete arcs whose intensity depends on the local geometry of charge separation and the specific atmospheric conditions at the moment of discharge, not simply on the peak field value. The induction ring at the mast base provides the complementary measurement needed to constrain this: it records the charge flux actually carried by dust passing through the ring cross-section, which together with the field mill record allows the electric field and charge flux measurements to jointly constrain the energy delivered by dust-driven electrostatic processes at each node over the mission duration. This joint constraint ,not the field mill record alone ,is what makes the chemistry comparison meaningful: when the final dataset shows that perchlorate concentration at depth at N1 is higher than at N2 by a specific margin, and the combined field mill and induction ring record shows that N1 received substantially greater cumulative electrostatic energy delivery than N2 over the measurement period, the causal connection between atmospheric forcing and surface chemistry can be evaluated quantitatively for the first time.

The four field mills together constitute a sparse but real spatial array of atmospheric electric field measurements. The correlation between the field records at the four nodes ,whether the dust devil signature appears simultaneously at all four or sequentially as the vortex traverses the array ,provides information about the spatial scale and tracking direction of individual vortices. This is a byproduct of the chemistry measurement, not its primary purpose, but it would be the highest-resolution ground-truth dataset on individual dust devil structure ever collected at a single Mars surface site.

The meteorological mast

The central node carries a 1.5 metre meteorological mast carrying instruments at three heights: 0.5 metres, 1.0 metres, and 1.2 metres, plus the electric field mill at the mast top. The mast instruments include an ultrasonic anemometer at 1.2 metres for wind speed and direction ,ultrasonic is specified because it has no moving parts and therefore no mechanical wear failure mode over a multi-year Mars surface mission ,temperature sensors at 0.5 and 1.0 metres for boundary layer thermal structure, a pressure transducer and relative humidity sensor at mid-mast height, a photodiode dust opacity sensor, and an induction ring charge sensor at the mast base that measures the electric charge carried per unit volume of dust passing through the ring cross-section during active dust events.

The induction ring provides a measurement distinct from and complementary to the field mill. The field mill measures the ambient electric field strength at its location. The induction ring measures the charge actually carried by dust particles passing through it during a vortex event. Together, the two measurements allow the relationship between field strength and charge density in Martian dust vortices to be characterised at the actual Mars surface, a relationship that has until now been estimated only from terrestrial analog work and numerical models. The mast mass estimate of four kilograms includes the post structure, all sensor heads, and cabling to the central node electronics package.

The inter-node ground-penetrating radar

The ground-penetrating radar system uses the central node at C0 as a transmitter and the three outer nodes as receivers. The transmit and receive antennae are coupled to the surface cable that already runs between the nodes for power and data, using the cable as a waveguide to carry the GPR signal between nodes without requiring a separate dedicated cable. This integration reduces the total cable mass and eliminates the need for a separately anchored GPR cable system.

The system operates in the 500 megahertz to three gigahertz frequency range, targeting subsurface imaging from the surface to approximately three metres depth. At these frequencies and at the expected dielectric properties of dry Martian regolith, the wavelength in the material is between five and thirty centimetres, giving vertical resolution of two to fifteen centimetres ,sufficient to detect layering at the scale of individual geological beds and at the scale of the depth-dependent oxidation profile the chemistry measurements are characterising.

The key engineering advantage of the fixed inter-node geometry over rover-mounted GPR is that the source-receiver separation is precisely known and does not change throughout the mission. Rover-mounted GPR ,as implemented on the RIMFAX instrument on Perseverance ,achieves its subsurface imaging by moving the receiver relative to the transmitter and using the changing geometry to compute a subsurface image. The computation requires accurate knowledge of rover position, and the resulting image resolution is limited by position uncertainty. The MGSS inter-node geometry is fixed from the moment of deployment, the source-receiver separation is measured once on sol zero, and every subsequent GPR measurement uses that same known geometry. The subsurface image between C0 and each outer node is therefore a persistent, repeated observation that can be averaged over the mission duration to achieve sensitivity below the single-pass noise floor.

The GPR image produced by the fixed array is a two-dimensional slice of the subsurface between the transmitter at C0 and each receiver node, extending to three metres depth over a lateral baseline of five to ten metres. Three such slices ,one per radial arm ,constitute a partial three-dimensional tomogram of the shallow subsurface in the MGSS deployment footprint. This tomogram is what allows the geological structure beneath the site to be separated from the atmospheric chemistry gradient in the interpretation of the chemistry measurements. If the GPR reveals a geological boundary ,a bed contact, a cemented layer, a lithological change ,that correlates with a step change in the chemistry profile, the geological boundary is the explanation for the chemistry pattern and not the atmospheric chemistry gradient. If the chemistry profile varies smoothly with depth and laterally across the array in a pattern consistent with the gradient framework, with no corresponding geological structure visible in the GPR image, the atmospheric chemistry gradient interpretation is strongly supported.

The sealed atmospheric exclusion chamber

The most original component of the MGSS is the sealed atmospheric exclusion chamber deployed at node N2. It is also the component most likely to draw reviewer scrutiny, because a sealed chamber that alters not only atmospheric access but also temperature, dust deposition, and surface boundary layer behaviour creates complications that must be accounted for rather than ignored.

The chamber is a cylindrical enclosure 300 millimetres in outer diameter with walls of three millimetre titanium alloy. An annular regolith skirt extends from the base of the chamber, driven fifty millimetres into the surface regolith by a spring-loaded actuator on deployment day, creating a continuous peripheral seal around the perimeter of the enclosed area. The skirt is hardened stainless steel with a sharpened lower edge to penetrate moderately consolidated regolith. If the surface at the N2 deployment location proves to be hard consolidated rock rather than loose regolith, the skirt will not penetrate and the chamber will not seal ,this is a real failure mode, and site selection must verify from lander camera imagery on sol zero that loose regolith is present at the intended N2 location before the actuator fires.

On deployment, a one-way check valve in the chamber wall allows the interior to equilibrate with the ambient Mars atmosphere over approximately twenty-four hours. After equilibration the valve closes, sealing the interior atmosphere at whatever the ambient pressure and composition were at that time. From that point, the interior of the chamber is isolated from the Martian atmosphere for the duration of the sealed period ,nominally one full Martian year, corresponding to 687 Earth days. The chamber removes atmospheric contact but does not remove galactic cosmic ray penetration. Radiolytic destruction of perchlorate by cosmic rays continues inside the sealed chamber at the same rate as outside it, because cosmic rays penetrate several metres of regolith and are unaffected by a thin metal enclosure. This is an important feature of the experimental design, not a flaw: the sealed node isolates atmospheric production while retaining radiolytic destruction pathways, meaning the difference in chemistry between the sealed and open nodes after one year measures the atmospheric production component alone, with the radiolytic destruction contribution held equal on both sides of the comparison. The controlled experiment is therefore a cleaner separation of the two competing processes than it would be if the chamber blocked cosmic rays.

The interior of the chamber is instrumented with a temperature sensor accurate to 0.1 Kelvin, a pressure transducer accurate to one pascal, and a relative humidity sensor. These instruments log continuously at one-minute resolution throughout the sealed period. The purpose of this logging is not to characterise the interior environment for its own sake but to provide the correction factors needed to interpret the chemistry comparison between the sealed and open nodes. The chamber interior will be warmer than the open surface during the day ,solar heating of the chamber structure raises the interior temperature relative to the ambient regolith ,and cooler at night, because the chamber insulates the enclosed regolith from radiative cooling. This thermal perturbation alters the rates of any chemistry that is temperature-dependent, including the radiolytic destruction of perchlorate and any residual moisture-driven reaction pathways. The temperature log quantifies this perturbation, and the correction is applied to the chemistry comparison explicitly, with the magnitude of the correction reported alongside the measurement so that reviewers can evaluate it.

The expected differential between the sealed and open node chemistry after one Martian year represents the integrated effect of one year of atmospheric processing at that specific site. If the triboelectric production mechanism is active at the rate the framework predicts, and if the radiolytic destruction is operating at the rate that cosmic ray penetration depths and perchlorate radiolysis cross-sections suggest, the sealed node should accumulate measurably less oxidant than the open nodes over the sealed period. The magnitude of the difference constrains both the production rate and the destruction rate independently, because the open nodes receive both processes and the sealed node receives only destruction. This is, in the strictest sense, a controlled experiment ,the first ever conducted on the Mars surface.

At the end of the sealed period, the chamber is vented by a second valve command, the interior pressure equilibrates with ambient, and the drill at N2 is deployed into the enclosed area. A port sealed with a lip gasket in the chamber base allows the drill auger to pass through while maintaining the interior in a condition representative of the sealed environment until drilling begins. The sonde is deployed into the N2 borehole immediately after drilling, before any significant re-equilibration of the interior chemistry with the now-open atmosphere. The resulting chemistry profile from N2 is compared against the profiles from N1 and N3, measured over the same period, at the same depths.

The two-scale phased architecture

The Pathfinder configuration just described is the first phase of a two-phase architecture. A second mission class, designated MGSS-Field, would extend the node geometry to a full logarithmic spoke pattern: three nodes at one metre from the centre along each radial arm, three nodes at ten metres, and three nodes at one hundred metres ,ten nodes total, sampling at three spatial scales simultaneously. This configuration would produce a true multi-scale spatial tomogram of subsurface chemistry, from the sub-metre scale at which individual rock clasts and pore structures control chemistry, through the metre scale at which the gradient framework predicts the steepest gradients, to the hundred-metre scale at which the full diameter of a small dust devil column is captured in a single array measurement.

The MGSS-Field architecture is not speculative science fiction. It is an engineering extrapolation from the Pathfinder, using the same components at larger scale, with the key addition that the hundred-metre baseline inter-node GPR requires a small dedicated surface rover traversing between the outer nodes rather than a fixed cable ,at that baseline, cable anchoring and signal integrity become engineering challenges that the fixed cable approach cannot cleanly solve. The two-sonde deployment required at Field scale, to service ten nodes in a reasonable cycling time, is a straightforward duplication of the single Pathfinder sonde. The mass envelope for MGSS-Field sits in the 150 to 200 kilogram range ,a medium lander primary payload, or a New Frontiers class mission dedicated to this measurement.

The phasing matters scientifically as well as programmatically. The Pathfinder result is designed to be decisive in a specific sense: it either confirms that a measurable chemistry gradient exists at the five to ten metre lateral scale and at the vertical scale from surface to one metre, or it does not. If it does not, MGSS-Field is not a logical next step ,the gradient framework requires revision before a larger investment is justified. If the Pathfinder confirms the gradient, MGSS-Field answers the follow-on question of how the gradient varies with spatial scale, which is the question needed to translate the framework into landing site selection criteria for biosignature missions. The two phases are connected by a conditional: the Field mission is scientifically justified if and only if the Pathfinder result is positive.

Mass budget and system integration

The Pathfinder mass estimate is built from component-level heritage where it exists and engineering judgement where it does not. The four drill units at a heritage-derived estimate of fifteen to eighteen kilograms each ,the TRIDENT flight unit is documented at sixteen kilograms in the VIPER configuration, with variation depending on the deployment mechanism and avionics integration ,contribute sixty to seventy-two kilograms to the total, with sixty-four kilograms used as the working estimate. This is the dominant mass driver and the figure most sensitive to the specific mechanical design of the fixed-position adaptation. The shared sonde system, including probe body, wireline cable, and deployment winch, is estimated at five kilograms. The meteorological mast at four kilograms. The four electric field mills at approximately 500 grams each, or two kilograms total. The inter-node ground-penetrating radar system, using the existing cables as waveguides rather than requiring dedicated antennae, is estimated at one kilogram for the transmitter and receiver electronics. The sealed atmospheric exclusion chamber at 1.5 kilograms. The central processor, power conditioning, and communications electronics at eight kilograms. Deployment hardware, anchoring, and cable management at five kilograms. The total landed mass estimate for MGSS-Pathfinder is approximately 68 kilograms. This estimate does not include the lander platform or the deployment mechanism that positions the nodes on the surface;those are lander system allocations, not MGSS allocations.

The power budget is driven by the drill units when active: each draws up to 200 watts peak during drilling operations. The drills are not operated simultaneously and are not in continuous operation ,the drilling campaign at each node is conducted over multiple Martian days, with each drill session consuming perhaps two to four hours of active drilling time. The continuous power draw from sensors ,four electric field mills, the meteorological mast suite, the GPR in listening mode, the electronics ,is estimated at 18 watts. A solar panel array sized for average insolation at the mission latitude can sustain this load during nominal operations, but continuous science return through regional dust storm events and Martian winter conditions ,both of which can reduce solar input by factors of two to five for periods of weeks ,requires RTG-class supplemental power. An RTG supplementing the solar array is therefore the baseline power architecture for a mission requiring uninterrupted sensor operation across a full Martian year, and the 18-watt continuous load is sized to be within RTG-only capability if the solar panels are entirely obscured. Drill operations are scheduled for periods of peak power availability and do not run during storm periods.

The technical risk register

A proposal of this kind is not credible unless its risks are named explicitly. There are five.

The first is borehole stability. The 1.7 millimetre clearance between the 22 millimetre sonde and the 25.4 millimetre borehole wall is adequate in consolidated material but may be insufficient in unconsolidated or mixed regolith, where partial wall collapse after drill withdrawal can prevent sonde entry. Two recovery options are available if initial sonde deployment meets resistance: a reaming pass with the drill at low torque and no percussion to clear collapsed material from the upper borehole, and vibration-assisted sonde descent using a small eccentric mass actuator on the sonde body to reduce wall friction during lowering. Neither recovery option guarantees success in severely unconsolidated material, so the primary mitigation remains site selection prioritising geologically consolidated terrain. Acceptance that sonde deployment may not be achievable at all four nodes is built into the minimum success criteria ,the minimum scientifically adequate dataset requires successful sonde deployment at two contrasting nodes, C0 as reference and one outer node for lateral comparison.

The second is sonde ion sensor performance. The ion-selective electrode package requires dissolved ionic species at the measurement point, which is not guaranteed in dry Mars regolith. If the ion sensors produce no signal, the Raman and VNIR measurements carry the chemistry dataset. The perchlorate Raman signature is well-characterised at concentrations above approximately 0.1 percent by weight, which is below the measured Mars surface concentration of 0.5 percent ,the Raman measurement is therefore the primary measurement even if the ion sensors fail.

The third is field mill qualification in CO_2 at Mars pressure. The Harrison and Marlton instrument was not designed or tested for this environment. The Paschen breakdown threshold in CO_2 at 700 pascals is roughly half that in air at sea level pressure for the same gap geometry. Motor insulation, electrode spacing, and housing gap geometry must all be reviewed against this threshold. This is a standard engineering analysis, not a research problem, but it has not been done because no field mill has previously been proposed for Mars surface deployment. The analysis should be completed before Phase A of any mission carrying the MGSS.

The fourth is sealed chamber thermal perturbation. The temperature differential between the interior and exterior of the sealed chamber may be larger than the 5 degree Celsius estimated above; if the chamber structure absorbs solar radiation more efficiently than expected ,a darkened or dusty chamber surface could warm significantly more than a clean one. If the interior-exterior temperature differential exceeds 10 degrees Celsius on average, the thermal correction to the chemistry comparison becomes large relative to the expected chemistry signal, and the controlled experiment interpretation is compromised. The mitigation is thermal modelling of the chamber under realistic Mars surface conditions before design finalisation, and specification of a surface finish that minimises solar absorption.

The fifth is the skirt deployment failure on hard ground. If the surface at N2 is consolidated rock rather than loose regolith, the spring-loaded skirt actuator will not penetrate and the chamber will not seal. No science is lost from the other three nodes if this occurs, but the controlled experiment ,the most scientifically original component of the MGSS ,will not be executable. The mitigation is pre-selection of the N2 deployment location from lander cameras before the actuator fires, with a backup N2 position designated in case the primary location proves unsuitable.

What the MGSS would confirm or refute

The measurement the MGSS makes has one of two outcomes, and both are productive. If the chemistry profiles at the open nodes show perchlorate and VNIR oxidation indicators decreasing with depth, and this depth profile correlates with the cumulative electric field delivery measured by the field mill at each node, and the sealed node accumulates measurably less oxidant than its open neighbours after one

Martian year, the framework presented in this book is confirmed at the spatial scale of the Pathfinder array. The confirmation does not prove the framework at all scales and all locations on Mars, no single field experiment can do that, but it establishes that the mechanism is real, that it operates at a measurable rate on a timescale accessible to a single mission, and that it varies laterally in a way that correlates with atmospheric forcing. That result would be sufficient to justify incorporating the gradient framework into landing site selection criteria for the next generation of biosignature missions.

If the open nodes show no depth-dependent chemistry gradient, no lateral variation that correlates with electric field forcing, and no measurable difference between the sealed and open nodes after one year, the framework in its present form is wrong at the spatial and temporal scales the Pathfinder probes. This is not a failure of the mission,it is a scientific result of equal value to a positive outcome. It means that either the spatial scale of gradient expression is larger than ten metres, that the active production rate is too slow to be measurable in one year, or that geological inheritance dominates over modern atmospheric processing at this site. Each of those alternatives is a specific, testable refinement of the framework that directs the next measurement rather than abandoning the question. A negative Pathfinder result would be the most precisely informative negative result in Mars surface science since Viking, because unlike Viking, the MGSS would have simultaneously recorded all the variables needed to interpret why the hypothesis failed, rather than leaving the failure open to competing explanations.

That is the standard to which a field experiment should be held. It should be designed so that both outcomes teach something specific. The MGSS meets that standard. Every Mars drilling mission that preceded it does not, because no preceding mission was designed as an experiment, each was designed as a measurement, and measurements have no outcome.

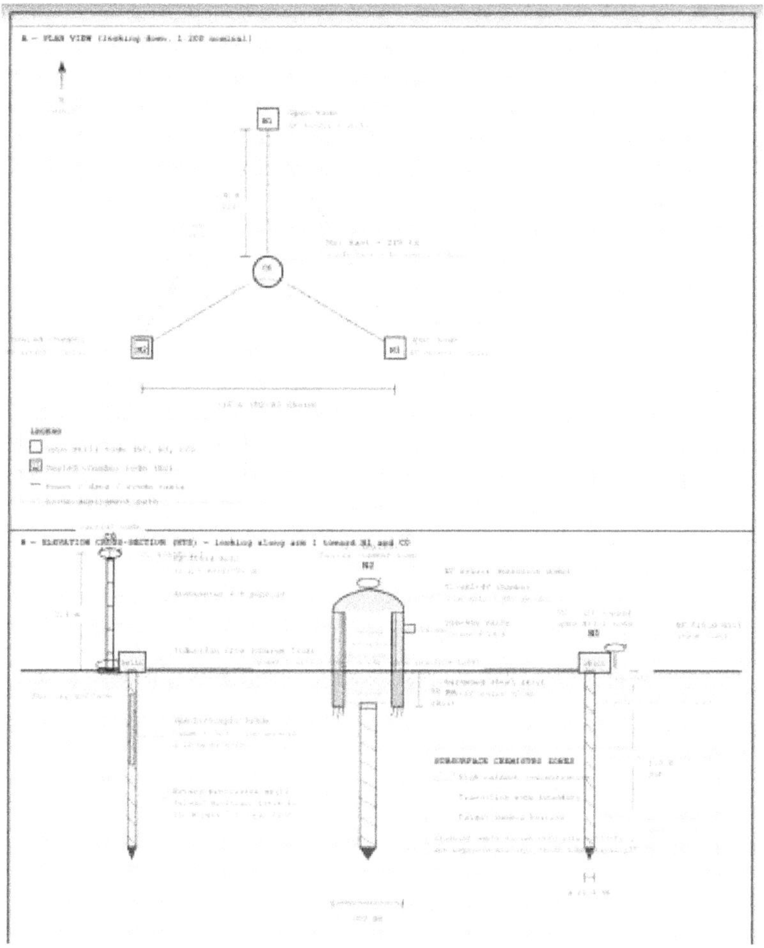

Figure 1. MGSS-Pathfinder system schematic - plan view (A) and elevation cross-section (B) of the MGSS-Pathfinder system. Node positions, radial spacing, and subsurface zone depths are nominal and schematic only. Actual deployment geometry will be determined by on-site assessment using lander imaging on sol 0. Subsurface chemistry zone boundaries are illustrative. Actual depth and transition sharpness depend on site dust activity history, regolith permeability, and exposure duration, and are among the primary quantities this instrument is designed to measure. All component dimensions are based on heritage estimates

and preliminary engineering; subject to revision during Phase A design and site characterisation. Scale: plan view approximately 1:200; elevation not to scale.

Conclusion: What the Chemistry Tells Us

The argument of this book can be stated in a single paragraph, though it took thirty-two chapters to develop properly. Mars has an active atmosphere that has been doing electrochemical work on its surface for as long as dust has been lofted into the thin Martian air. That work produces oxidants, concentrates hygroscopic salts, charges mineral particles, and modifies whatever ancient chemistry was originally present in exposed rock. The result is a planet whose surface records not one history but two , the ancient geological history that astrobiologists have spent decades reading, and a modern atmospheric processing history that has been editing that record continuously since each rock was first exposed. Neither history can be understood in isolation from the other. That interaction is what this book has attempted to map.

The evidence for the modern processing is not circumstantial. Phoenix detected perchlorate at half a percent by weight in polar soil. Curiosity found comparable concentrations thousands of kilometres away in a completely different geological setting. Laboratory work identified the triboelectric discharge mechanism and demonstrated that it produces perchlorate at rates orders of magnitude faster than ultraviolet photochemistry under Mars-relevant conditions. In October 2024, the SuperCam microphone aboard Perseverance recorded fifty-five discrete electrical discharge events in Martian dust devils and storm fronts , the direct acoustic signature of the chemistry this book describes, captured in real time on the surface of another planet. The mechanism is confirmed. What remains uncertain is its spatial structure, its rate at specific locations, and the depth to which its effects penetrate the regolith. Those are the measurements this book has argued for throughout.

The scientific implications follow directly from the mechanism. A biosignature is only present in a sample if the sample has preserved it, and a sample has only preserved it if the oxidative processing that has been running on exposed Martian rock for billions of years has not destroyed it. This means that ancient habitability and preservation quality are different variables, and that treating them as equivalent , as mission planning has largely done by selecting sites for ancient geological quality and assuming preservation follows , is the source of the ambiguity in the biosignature record accumulated to date. The sites we have sampled are not bad science. They are good science conducted without a variable that turns out to matter. Adding that variable , the modern chemistry gradient, characterised by dust activity history, surface age, and sample depth , does not invalidate what has been found. It sharpens the interpretation of negative results and increases the probability that future missions sample material that actually preserves what they are looking for.

The engineering implications are equally direct, and they were the more surprising part of this project to develop. The same atmospheric process that complicates biosignature preservation is simultaneously a resource production system. Mars concentrates perchlorate through triboelectric chemistry, and that perchlorate is an oxygen feedstock. It concentrates hygroscopic salts, and those salts are a water-recovery pathway. It electrostatically charges and size-sorts dust particles, and that pre-sorted charged feedstock can be intercepted by a collector array rather than treated as contamination to be excluded. The electrochemical architecture described in Part Two and Part Three is not a scheme for making the best of a difficult situation. It is a recognition that Mars has been doing industrial chemistry for billions of years using solar energy as its power source, and that a settlement which understands and interfaces with that chemistry has access to resources that a settlement treating the surface as passive rock does not.

The nitrogen gap identified in Chapter 29 is the honest limit of that optimism. Oxygen from atmospheric CO_2 electrolysis is solved at TRL 9. Water from subsurface ice is demonstrated at scale on Earth. Power from fission surface reactors is approaching flight readiness. But nitrogen , the diluent gas that any human habitat requires to prevent oxygen toxicity and eliminate fire risk , is present in the Martian atmosphere at two and a half percent, a concentration that requires a five-hundred-fold enrichment to reach habitat partial pressures. The plasma nitrogen fixation work from Antwerp and Lisbon is genuinely promising, and the connection between engineered plasma reactors and the triboelectric plasma chemistry this book describes throughout is not coincidental , both exploit the same physical fact that Mars surface pressure sits at the ideal operating point for non-thermal plasma reactions. But plasma nitrogen fixation for Mars is at TRL 3, not TRL 9, and the difference between those two numbers is the distance between a laboratory demonstration and a system a crew can depend on. That distance has to be covered before the first human mission, not after it.

The Mars Gradient Sampling System, described in Chapter 32, is where the science and the engineering converge into a single testable question. The MGSS is not a resource extraction system. It is an instrument designed to determine whether the scientific framework of this book is correct , whether the oxidation gradient is real, spatially structured, correlated with atmospheric forcing, and actively maintained at a measurable rate. Its four-node geometry, its coupled chemistry and electric field measurements, its inter-node GPR tomography, and its sealed atmospheric exclusion chamber together constitute the first Mars surface experiment specifically designed to test a planetary process rather than characterise a sample. If it confirms the gradient, every argument in Parts One through Three is strengthened and the case for incorporating atmospheric chemistry into mission planning becomes empirical rather than inferential. If it refutes the gradient at the spatial scale it probes, the framework requires revision and the revision is guided by the same dataset that identified the failure. Either way, the question moves forward.

What has not moved forward, and what this book argues must change, is the underlying philosophy of how Mars missions select their targets. The current approach optimises for ancient geological quality , the most compelling ancient environment, the best-preserved stratigraphy, the strongest mineralogical evidence for past water. Those are necessary conditions for finding biosignatures but they are not sufficient ones. The sufficient condition is that the ancient quality is matched by modern preservation quality , that the site's atmospheric activity history has not destroyed what its geological history created. That second condition requires measuring the modern chemistry, characterising the gradient, and selecting sites where the two factors align. That measurement does not require new physics. It requires the MGSS, or something functionally equivalent, to fly before the next generation of biosignature missions commits to a landing site.

The same argument applies to the engineering missions. The electrochemical architecture described in Part Three is a framework for sites with moderate to high dust activity , sites where the feedstock delivery rate justifies the mass of collection and processing hardware. Site selection for the first human Mars missions is currently conducted without dust activity index as a formal parameter. It should not be. The data exists in the Context Camera archive. The analysis is straightforward. The parameter captures information that currently exists in no site selection scoring scheme. Adding it costs nothing except the recognition that modern atmospheric chemistry is as relevant to engineering as it is to science , which is the central claim of this book.

Fifty years after Viking found reactive soil on Mars and interpreted it as a legacy of ancient processes, it is possible to say with confidence that the soil is reactive because Mars is actively processing it now, has been doing so continuously, and will continue doing so. The reactive species are not ancient. The planet is not a museum. The chemistry that frustrated the Viking biology experiments in 1976 is the same chemistry that produces the oxygen feedstock a Mars settlement will extract, the same chemistry that controls which ancient samples have preserved their biosignature content and which have not, and the same chemistry that charged the dust in the dust devil that crossed over Perseverance in October 2024 and crackled in the microphone. That continuity is the point. The Martian surface connects 1976 to the present and forward to whenever the first human crew lands. Understanding it as a connected system , not compartmentalised into ancient geology and modern chemistry and future engineering, but integrated across all three , is what this book has attempted to provide. Whether the framework holds in its details will be determined by measurements that have not yet been made. The measurements are specified. The hardware exists in concept and in heritage. What remains is the decision to make them.

References

Hecht, M. H., Kounaves, S. P., Quinn, R. C., West, S. J., Young, S. M. M., Ming, D. W., Catling, D. C., Clark, B. C., Boynton, W. V., Hoffman, J., DeFlores, L. P., Gospodinova, K., Kapit, J., & Smith, P. H. (2009).

Detection of perchlorate and the soluble chemistry of Martian soil at the Phoenix lander site. *Science, 325*(5936), 64–67.https://doi.org/10.1126/science.1172466

Glavin, D. P., Freissinet, C., Miller, K. E., Eigenbrode, J. L., Brunner, A. E., Buch, A., Sutter, B., Archer, P. D., Atreya, S. K., Brinckerhoff, W. B., Cabane, M., Coll, P., Conrad, P. G., Coscia, D., Dworkin, J. P., Franz, H. B., Lashley, K. A., Leshin, L. A., Mahaffy, P. R., ... Steele, A. (2013).

Evidence for perchlorates and the origin of Martian organic compounds. *Journal of Geophysical Research: Planets, 118*(10), 1955–1973.https://doi.org/10.1002/jgre.20144

Carrier, B. L., & Kounaves, S. P. (2015).

The origins of perchlorate in the Martian soil. *Geophysical Research Letters, 42*(10), 3739–3745.https://doi.org/10.1002/2015GL064290

Atreya, S. K., Mahaffy, P. R., & Wong, A.-S. (2011).

Methane and related trace species on Mars: Origin, loss, implications for life, and habitability. *Planetary and Space Science, 59*(2–3), 133–136.https://doi.org/10.1016/j.pss.2010.08.016

Jackson, W. A., Böhlke, J. K., Andraski, B. J., Fahlquist, L., Bexfield, L., Eckardt, F. D., Gates, J. B., & Davila, A. F. (2015).

Global patterns and environmental controls of perchlorate and nitrate in arid environments. *Geochimica et Cosmochimica Acta, 164*, 502–522.https://doi.org/10.1016/j.gca.2015.03.016

Krauss, C. E., Horányi, M., & Robertson, S. (2003).

Experimental evidence for electrostatic discharging of dust near the surface of Mars. *New Journal of Physics, 5*, 70.https://doi.org/10.1088/1367-2630/5/1/370

Kok, J. F., & Renno, N. O. (2009).

Electrostatics in wind-blown sand. *Physical Review Letters, 102*(13), 138001.https://doi.org/10.1103/PhysRevLett.102.138001

Renno, N. O., Abreu, V. J., Koch, J., Smith, P. H., Hartogensis, O. K., De Bruin, H. A. R., Burose, D., Delory, G. T., Farrell, W. M., Watts, C. J., Garatuza, J., Parker, M., & Carswell, A. (2004).

MATADOR 2002: A pilot field experiment on convective plumes and dust devils. *Journal of Geophysical Research: Atmospheres, 109*(D16), D16203.https://doi.org/10.1029/2003JD004362

Farrell, W. M., Delory, G. T., Marshall, J. R., Cummer, S. A., & Renno, N. O. (2004).

Electric and magnetic signatures of dust devils from the MATADOR desert tests. *Journal of Geophysical Research: Planets, 109*(E3), E03004.https://doi.org/10.1029/2003JE002088

Farley, K. A., Williford, K. H., Stack, K. M., Bhartia, R., Chen, A., de la Torre Juárez, M., et al. (2020).

Mars 2020 mission overview. *Space Science Reviews, 216*, 142.https://doi.org/10.1007/s11214-020-00762-y

Wiens, R. C., Maurice, S., Robinson, S. H., Nelson, A. E., Cais, P., Bernardi, P., et al. (2021).

The SuperCam instrument suite on the Mars 2020 rover. *Space Science Reviews, 217*, 4.https://doi.org/10.1007/s11214-020-00777-5

Dartnell, L. R., Desorgher, L., Ward, J. M., & Coates, A. J. (2007).

Modelling the surface and subsurface Martian radiation environment: Implications for astrobiology. *Geophysical Research Letters, 34*, L02207.https://doi.org/10.1029/2006GL027494

Murchie, S. L., Arvidson, R. E., Bedini, P., Beisser, K., Bibring, J.-P., Bishop, J. L., et al. (2009).

Compact Reconnaissance Imaging Spectrometer for Mars (CRISM) on Mars Reconnaissance Orbiter. *Journal of Geophysical Research, 114*, E00D07.https://doi.org/10.1029/2009JE003344

Eigenbrode, J. L., Summons, R. E., Steele, A., Freissinet, C., Millan, M., Navarro-González, R., et al. (2018).

Organic matter preserved in 3-billion-year-old mudstones at Gale crater, Mars. *Science, 360*(6393), 1096–1101.https://doi.org/10.1126/science.aas9185

Navarro-González, R., Vargas, E., de la Rosa, J., Raga, A. C., & McKay, C. P. (2010).

Reanalysis of the Viking results suggests perchlorate and organics at mid-latitudes on Mars. *Journal of Geophysical Research, 115*, E12010.https://doi.org/10.1029/2010JE003599

Encrenaz, T., Greathouse, T. K., Lefèvre, F., Atreya, S. K., Bézard, B., Forget, F., & Smith, M. D. (2012).

Hydrogen peroxide on Mars: Observations, modeling, and implications. *Planetary and Space Science, 68*(1), 3–17.https://doi.org/10.1016/j.pss.2011.12.014

www.ingramcontent.com/pod-product-compliance
Lightning Source LLC
LaVergne TN
LVHW052320200926
848481LV00002B/616
* 9 7 9 8 2 3 3 5 4 9 3 1 1 *